AF342484

Efficient Algorithms of Time Series Processing and their Applications

EFFICIENT ALGORITHMS OF TIME SERIES PROCESSING AND THEIR APPLICATIONS

G. SH. TSITSIASHVILI

EDITOR

Nova Science Publishers, Inc.

New York

NOTICE TO THE READER

LIBRARY OF CONGRESS CATALOGING-IN-PUBLICATION DATA

ISBN: 978-1-60692-062-6
Available upon request

Published by Nova Science Publishers, Inc. ✦ New York

CONTENTS

PREFACE

The given collected articles presents works on processing time series of observations in problems of meteorology, ichthyology, medical geography, epidemiology and demography. These works have been published by authors within the last 4 years in the Russian journals and reported at various Russian and international conferences. The basic methods of processing of time series in the collected works are the developed algorithms for:

- recognition of images,
- classifications,
- estimations of dispersions of fluctuations concerning a trend.

The idea of construction of the first two algorithms consists in studying large outliers in a time series. Such an approach has allowed to construct quite simple for understanding and rather fast, as to computing, algorithms of recognition of images and classifications and to apply them in the problems that are characterized by large volumes of empirical information.

The third of the specified algorithms is based on special transformations of time series to problems with a small trend and greater fluctuations. Application of traditional algorithms in the considered arrays of the empirical information demands complex calculations. The problems described in the presented works are actual and that's why the use of the offered algorithms carries not an illustrative, but a substantial character. The problems in question:

- influence of meteorological factors on critical values: catch of fish (hunchback salmon) in the Amur River, freezing in Tatar Strait, numbers infected by tick-borne [vernal] encephalitis and other epidemic diseases in the Primorye Territory,
- influence of economic transformations on various age groups of the population and on dynamics of a population in cities of Primorye Territory,
- influence of global warming on fluctuations of surface temperature in various areas of the Far East.

This book is needed because a lot of new applied problems in concrete research areas originated in recent years. These problems are connected with complicated economical transformations and reforms, global climate warming, etc. Traditional methods of modeling and data processing can not solve them because of their large complexity. So it is necessary to extract specifics of these problems which largely changes of the main characteristics of considered systems. The suggested collection of papers contains different approaches of extracted specifics application in data processing and next analysis. These approaches allow one to construct efficient data processing algorithms and to simplify significantly analysis of considered systems.

In chapter 1, a new approach to the choice of information indicators of climate (atmospheric pressure near the earth surface, in the middle troposphere and near-ground air temperature) that is based on visual estimation of intra-annual bundle fluctuations of climatic parameter trajectories is proposed. Disclosure of intra-annual bundles of climatic parameters provided the opportunity to develop a unique algorithm for the extreme conditions recognition. The possibility to forecast dramatic events has been studied over dynamics of the attendant feature. Operation of the proposed algorithm is described using the example of recognition of "critical" levels of the Asian pink salmon catch in different fishery regions in the Far Eastern seas. It is concluded that the suggested method can be used for the forecast of extreme conditions in the natural processes.

In chapter 2, the peculiarities of atmospheric processes in the middle troposphere that stimulate formation of anomalous ice cover in the Tatar Strait were explored. A clear distinction was discovered in AT_{500} fields above the central sector of the second natural synoptic region in the Northern Hemisphere under small and large ice cover of the aquatic area of the northern part of the Japan sea. For the experimental prognosis of extreme ice cover in the Tatar Strait was used the method of interval recognition was used. Analysis of the coefficients of recognition according to data of above-earth pressure and H500 magnitudes for maximal and minimal ice cover in the Tatar Strait exhibited that summer (June, July) and autumn (September, October) months possess the most distinguished recognition values that is in accordance with the physical explanation of possible causes.

In chapter 3, to analyze peculiarities of surface air temperature the authors used least square method and find as trend coefficient so least square. Then they calculated these characteristics for temperature dynamics of different meteorological stations.

In chapter 4, the special epidemic prognosis of the tick-borne encephalitis foci functioning was tested. The temporal extrapolation prognosis of tick-borne encephalitis morbidity in the Primorye area is made by means of autocorreclation models. A novel technique of the temporal factor prognosis of the tick-borne encephalitis infection has been worked out and tested both from the retrospective review and in real time.

In chapter 5, a technique of temporal factor prognosis of critical levels of infection diseases has been elaborated and tested. This approach has a methodological and applied importance and may be used in medico-ecological and epidemiological investigation.

In chapter 6, in conditions of unstable demographic indicators used as a basis for administrative decisions, the most difficult thing is definition of their orientation. The offered analysis of network integrated schedules of passage of various generations through an economy changing in time has shown, that there are branches in the demographic system which are most unstable to external pressure.

In chapter 7, the article is devoted to analysis, performed by means of integrated operational schedule, of population changes in the cities of the Russian Far-East with various population in time domain containing a crucial moment of development. The graphic analysis with the different size of scales has led to delimiting aggregated dynamics on two clusters– the cities with decreasing and with stable population. Two main factors that have a positive influence on the dynamics of the development of cities were clearly recognized. They are –

frontier location and presence of the government enterprises, which are extremely important for the Russian state.

In chapter 8, a problem of estimation of a random deviation between observations and a regression function is considered. A problem of a variation estimation usually is solved by an empirical variation. But even for this widely used statistic, it is complicated to calculate its own variation. To calculate a variation of an estimate for a variation of a deviation from a polynomial regression function is much more difficult. Nevertheless it is possible to solve this problem if observations are made in integer-valued points. The problem solution is represented in this chapter.

This suggested collection of papers may be interesting for specialists in data processing and for specialists in concrete subject areas: epidemiology, meteorology, fishing, medical geography, etc. As for my part, young people (students and post graduate students) may use suggested algorithms in different listed applications.

While this suggested collection of papers makes it possible to simplify the data analysis procedure, its main advantage is in the close connection of mathematical and algorithmic approaches with considered problems and with specifics of their functioning. Moreover all considered problems are actual in concrete areas of research and their main similarity is in the presence of large changes of systems characteristics.

Chapter 1

INTRA-ANNUAL BUNDLES OF CLIMATIC PARAMETERS

V.A. Svyatukha, G.Sh. Tsitsiashvili[1],* T.A. Shatilina[2],** and A.A. Goryainov***

* Institute of Applied Mathematics, FEB RAS, Vladivostok, Russia,
** Pacific Research Fisheries Center, Vladivostok, Russia,

ABSTRACT

A new approach to the choice of information indicators of climate (atmospheric pressure near the earth surface, in the middle troposphere and near-ground air temperature) that is based on visual estimation of intra-annual bundle fluctuations of climatic parameter trajectories is proposed. Disclosure of intra-annual bundles of climatic parameters provided the opportunity to develop a unique algorithm for the extreme conditions recognition.

The possibility to forecast dramatic events has been studied over the dynamics of the attendant feature. Operation of the proposed algorithm is described usimg the example of recognition of "critical" levels of the Asian pink salmon catch in different fishery regions in the Far Eastern seas.

INTRODUCTION

Regressive models describing mainly the averaged dynamics, and not the skips (the term of random process theory) over a high level, are the general mathematical methods that are used to forecast hydrometeorological phenomena in the Russia Far East. Statistic processing of temporal series is fulfilled by the standard techniques that have a few defects. They are not capable of revealing critical moments (anomalies) and natural cycles clearly enough (possibly, wrong). Almost all the forecasts based on the correlation between environment

---

[1] guram@iam.dvo.ru.
[2] shatilina@tinro.ru.

parameters are of low reliability (not more than 50–60%). Especially if it concerns the forecast of anomalous (disastrous) hydrometeorological processes that produce great economic damage.

As a result, the necessity to use other mathematical methods for the forecast of extreme climatic events is in urgent demand.

The present work offers a new approach to the choice of climatic indicators (atmospheric pressure near the Earth's surface and in the middle troposphere and surface air temperature) that is based on visual estimation of fluctuations in the intra-annual bundles of climatic parameter trajectories.

1. DATA

Average monthly data on the surface atmospheric pressure in the points of regular 5° grid for 1960–2000 placed at the web page of NOAA-CIRES Climate Diagnostic Center *http://www.cdc.noaa/gov/PublicData* and geopotential H_{500} at 19 aeroalogical stations of the Russia Far East located along the perimeter of the Japan and Okhotsk Seas and the Kuril Islands during 1950–1989 have been used as initial data. Additionally, data on the surface air temperature in Vladivostok and Terney provided by Primorsky Hydromet office for 1942–1992 and data on the catch of Asian hunchback salmon during 1950–1989 have been gathered (Yanovskaya et al., 1989; Summary on salmon fishery catch, 1995, 1998, 2000).

Analysis of interannual dynamics of climatic parameters testifies to their chaotic character that is hardly subject to the correct statistic analysis (since the temporal series is very short). The average annual change of surface pressure in Vladivostok is shown in Figure 1 as an example.

Analysis of intra-annual change of climatic parameters turned out to be more interesting and significant for their mathematic interpretation and development of extreme phenomena forecasting techniques.

Thus, Figure 2 shows variability of H_{500} geopotential above Vladivostok and Okhotsk during the course of the year for the long period (1950–1989). It is clear that intra-annual H_{500} change for the long-term period forms a distinct bundle of trajectories, i.e. has a small scatter of magnitudes (about 5 dkm). Intra-annual change of H_{500} geopotential at all the aerological stations, including offshore ones, has the same features. It is evident that intra-annual dynamics of average monthly values of geopotential H_{500} has a clearly defined trend at the background of small fluctuations (their amplitude is one order less than the trend amplitude). It can be concluded for sure that intra-annual dynamics of average monthly H_{500} values is repeated every year and produces rather a distinct bundle as a result. Statistical data samples increase from 40 (interannual dynamics) to 480 (for all months of a year).

Intra-annual dynamics of the near-ground pressure field has the same type only above the continental regions of Asia and coast of the Sea of Japan, i.e., above the regions of clear exposure of the Far East monsoon. It should be noted also that intra-annual run of atmospheric pressure near the surface and in the middle troposphere (at a 5000 m height) is in anti-phase, which is explained by widely known regularities of the synoptic processes evolution near the surface and at a certain height. Unlike the intra-annual change of H_{500} geopotential, the intra-annual run of near-ground pressure above the seawater areas varies significantly and does not form a distinct resultant bundle.

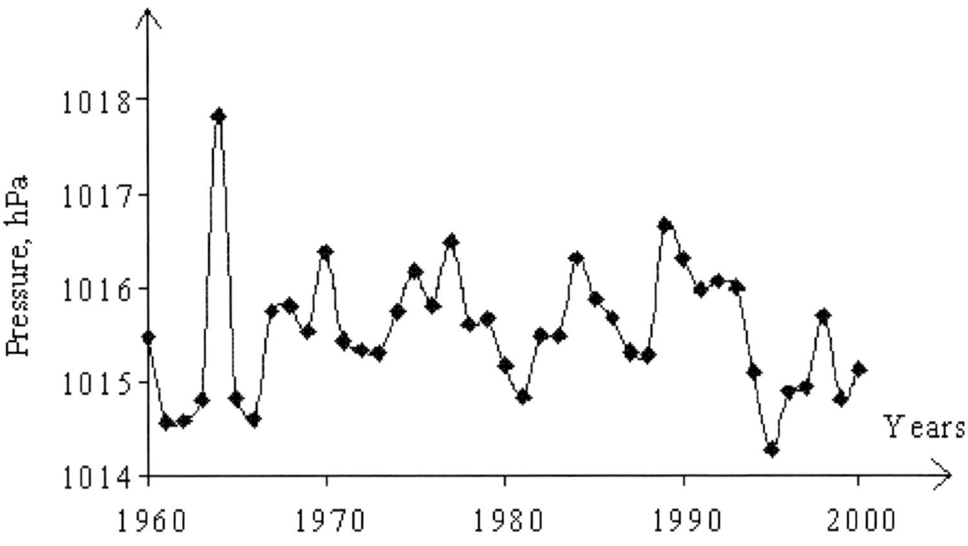

Figure 1. Average annual run of near-ground pressure in Vladivostok during 1960–2000.

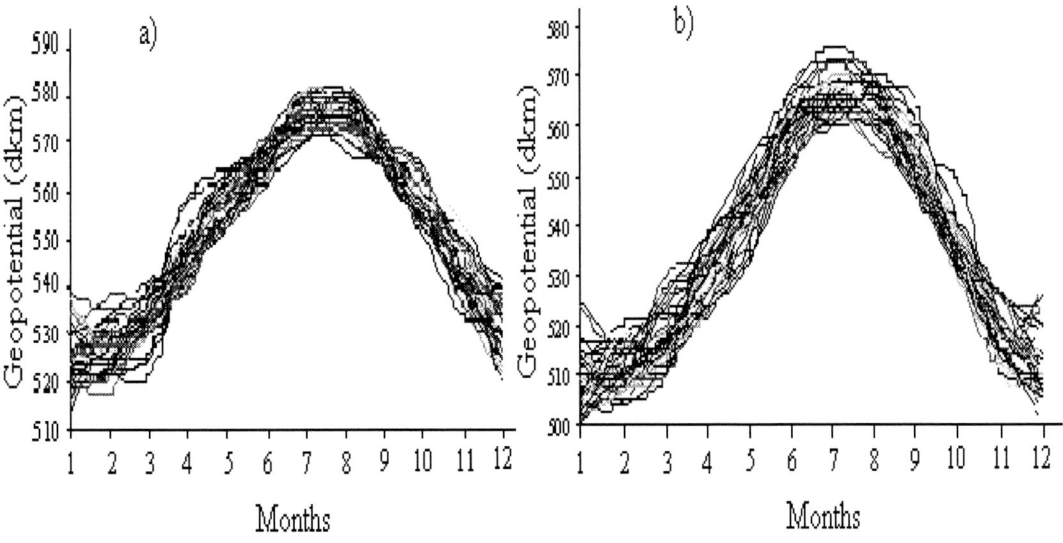

Figure 2. Intra-annual dynamics of H500 geopotential during 1950–1989 in: a) Vladivostok, b) Okhotsk.

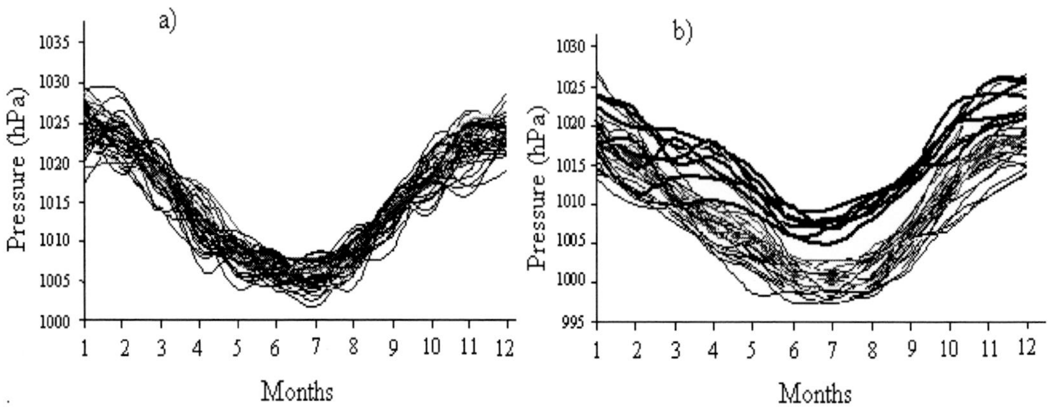

Figure 3. Intra-annual run of near-ground pressure during 1960–2000 in: a) Vladivostok, b) center of South-Asian depression (30°N, 90°E).

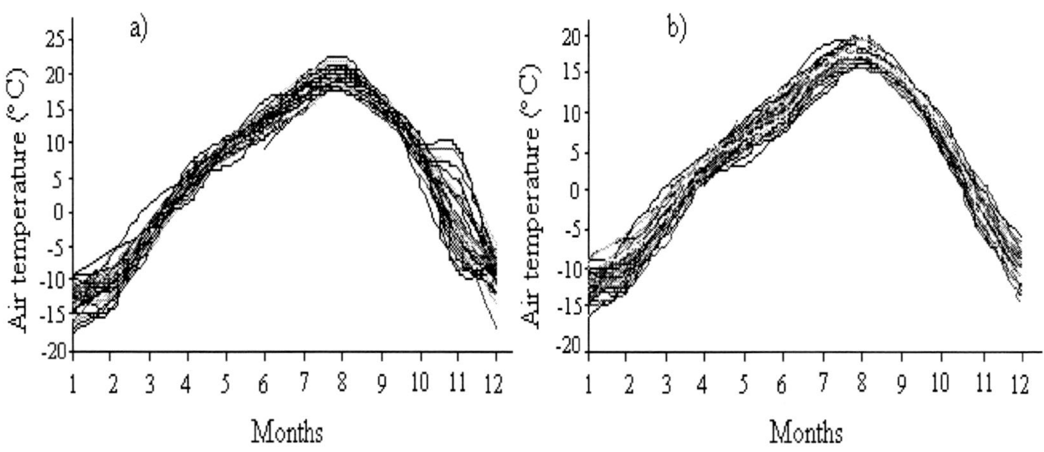

Figure 4. Intra-annual run of air temperature during 1942–1990 in: a) Vladivostok, b) Terney.

Figure 3a shows an example of intra-annual run of near-ground pressure above Vladivostok during 1960–2000. It is evident that the width of this bundle is within 5 hPa (the feature, which has a small scatter). Figure 3b shows the intra-annual run of near-ground pressure in the center of Asian depression (30°N, 90°E) during the same period. Two periods are distinguished that produce different levels of near-ground pressure, namely, 1960–1993 with low pressure and 1994–2000 (bold lines) with high pressure. It also should be marked that two different bundles are distinctly distinguished in the center of Asian depression (center of action in atmosphere) only. It possibly means that at the edge of 1993–1994 the epoch climatic changes occurred.

The distinct bundles of intra-annual run of near-ground air temperature, which is a thermal indicator of climatic changes, have been found. Figure 4 shows an example of intra-annual run of near-ground air temperature in Vladivostok and Terney (North Primorye) during 1942–1992. It is remarkable that intra-annual change of near-ground air temperature coincides with intra-annual change of H_{500} geopotential that is a dynamic parameter of atmosphere. It is also known that the middle troposphere is an energetic level of atmosphere.

2. METHODS INVESTIGATION

Revealing intra-annual bundles in the climatic parameters (H_{500} geopotential, near-ground pressure, air temperature) gave a chance to develop the algorithm for the extreme situations recognition. The possibility to forecast dramatic events has been investigated according to the dynamics of indicators of a coupled feature. Figure 5 shows the algorithm of the decisive rule development for recognition of dramatic events. Upper diagram demonstrates dynamics of the main feature that is subject to a high intra-annual variability (ice cover of seas, catch of Asian hunchback salmon in the local areas, precipitation, etc). When choosing the critical level that is specific for the main feature, we mark with "x" sign those moments of time, when dynamics trajectory of the main feature exceeds the established level (i.e., the anomaly moments of deviation). In our calculations the frequency of the exceeded criterion level was no more than 10%. The rest of moments of time are marked by "0" sign. Then we analyze dynamics of the attendant feature that is characterized by the small scatter of intra-annual variations (pressure on isobaric surface 500 hPa, near ground pressure and air temperature, etc). Then the magnitudes of the attendant feature during "dramatic" moments of time (marked by "x") are projected to the axis of ordinates and minimum B and maximum A values of the attendant feature are found. The next step is to develop a rule of recognition of a certain moment of time belonging to the critical level according to the belonging of the attendant feature to a section *[A, B]*. If during a non-dramatic moment of time the attendant feature is found in the section *[A, B]*, then the given moment will be mistakenly attributed to the critical one. This procedure is generalized by finding a separate critical section for each attendant feature. Thus, belonging of a certain moment of time to the critical one will be identified according to the belonging of each of attendant features to their own critical sections.

This task satisfactorily meets demands for decision within the following model on the images recognition.

The algorithm of recognition of extreme events was applied to forecast extreme ice cover in the Okhotsk Sea (Tsitsiashvili et al., 2002).

May the curves $X(t) = (X_j(t-1),\ j=1,...,12)$ of average monthly dynamics of H_{500} geopotential during a year $t-1$ that is proceeding to the year t from a set of observed years T be the objects and the expression $\{X(t),\ t \in T_1\}$, $\{X(t),\ t \in T_2\}$, where T_1 is the combination of years with maximum (minimum) ice cover $T_1 = T_2$ – addition T_1 to T_2 be the classes of those objects.

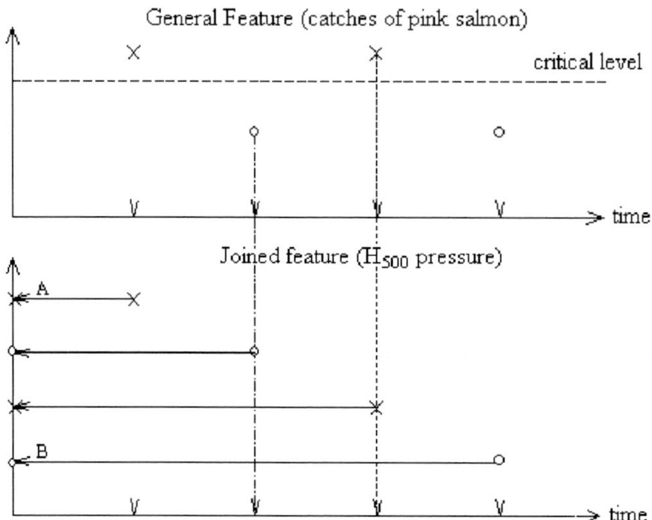

Figure 5. Construction of lines [A, B] recognizing excess by general feature of critical level.

At a first glance, such a formulation of task looks traditional for the images recognition, however, a new element is introduced here consisting of the class of extreme (maximum or minimum ice area) years.

The peculiarity of a new method is the choice of the following condition as the simplest decision rule for attribution of an object $(y_j, j = 1,...,12)$ to class T_1:

$$a_j \le y_j \le b_j, j = 1,...,12,\ \ \ \ \ \ \ \ \ \ \ \ \ \ (1)$$

where: $a_j = \min(x_j(t-1), t \in T_1)$, $b_j = \max(x_j(t-1), t \in T_1)$. In other terms, the object $(y_1,...,y_{12})$ is attributed to T_1 class, if the corresponding trajectory of average monthly dynamics during the course of the year gets within the bundle of trajectories corresponding to pre-extreme years. This version of the decision rule is unique for recognition of extreme environmental conditions by the data available.

The given decision rule is based on the ideas of interval mathematics and can be very quickly calculated by a computer. The quality of the proposed decision rule will be characterized by the rate of mistakes of the first and second type within the initial data sample $\{X(t), t \in T\}$.

A mistake of the first (second) type means a wrong identification of extreme (non-extreme) year in the initial data sample. Proposed decision rule provides the estimation of frequency of mistaken identification guaranteed from data omission (Bolotin and Tsitsiashvili, 2003).

3. DISCUSSION OF RESULTS

Our calculations based on the 40-year data on H_{500} geopotential with 4 years of anomalously high ice cover in the Okhotsk Sea as observed at 5 stations, namely, Okhotsk, Petropavlosk-Kamchatsky, Blagoveshensk, Kharbin, Tateno (Japan), have shown that this frequency at every station equals zero. It means that it is possible to introduce a summarized decision rule for the identification of years of maximum ice formation that would consist of the observance of inequality (1) by data obtained from at least one of 5 chosen stations. Here, the rate of mistakes of the first and second type will again be equal to zero.

So, if the average monthly climatic parameters (near ground pressure and air temperature, geopotential) during one year (two years) belong to the trajectory bundles shown in Figures 2–4, then the same features will belong to the narrower bundles during the critical years. The fact of belonging of the curve of intra-annual run of average monthly parameters to the narrow bundles is just a condition for the forecast of dramatic events.

The main entity of salmon fishery in the Russia Far East is pink salmon, which share takes no less than 40% of the total catch (Shuntov, 1994). This species has been thoroughly studied for a long time, however, accurate forecasting of salmon catch faces a regular failure in practice every year. One of the reasons of poor fishery forecasts (and not for salmon only) is linked to the fact that dynamics of pink salmon catch is characterized by a strong or chaotic variability, namely, it is greatly different from the intra-annual run of the above-mentioned climatic parameters (being a feature undergoing significant scatter in intra-annual variability). For example, pink salmon catch near western Kamchatka can fluctuate from several hundreds to tens of thousand tons.

Let us consider the operation of our proposed unique algorithm for recognition of a "critical" level in the Asian salmon catch in different fishery regions of the Far Eastern seas according to the H_{500} geopotential data (other climatic parameters were not considered in the work).

Figure 6 shows results of the quality of the "critical" level recognition in the catch of Asian pink salmon in the fishery regions of the Far Eastern seas (North-Okhotsk, Amur, Primorye, East Kamchatka, West Kamchatka, East Sakhalin, South Kurils) according to the H_{500} geopotential data from 19 aerological stations of the Russia Far East placed along the perimeter of the Japan and Okhotsk Seas (which characterize climate above the fishery regions) during January–December of 1950–1989. Each station of a separate year is characterized by the set of average monthly values of H_{500}.

Y axis shows the number of stations, where 100% correct recognition of excess of the assigned level of pink salmon catch according to data on H_{500} geopotential during two previous years is found.

For example, in West Kamchatka the number of stations with 100% correct recognition of the salmon catch exceeding 22,000 tons amounts to 18 (almost all stations), the number of critical years is equal to 8. In case of the pink salmon catch exceeding 32,000 tons, all 19 stations produce 100% correct recognition (figure 6b), the number of critical years is 5.

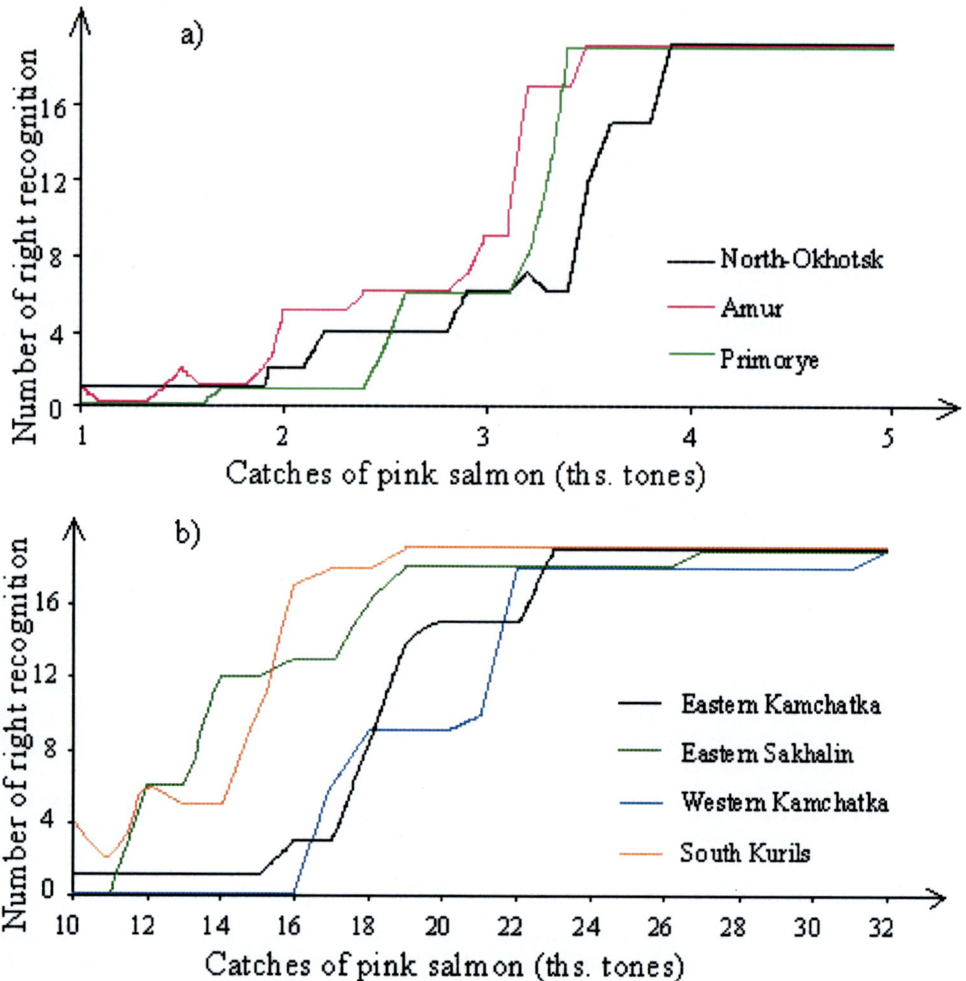

Figure 6. Quality of recognition of a critical level in the pink salmon catches over different Far Eastern fishery regions dependence on the number of right recognition at aerological stations of H500.

CONCLUSION

Disclosure of intra-annual bundles of some climatic parameters gave the opportunity to develop a unique algorithm for recognition of extreme natural phenomena. The possibility to forecast disastrous phenomena (a feature possessing the great scatter of intra-annual variations) by the dynamics of the attendant feature with a small scatter of values was shown to be possible.

Application of the proposed algorithm for recognition of "critical" levels in the Asian pink salmon catch (during two precedent years) and ice cover in the Okhotsk Sea has shown that the method proposed can be used to forecast extreme hydrometeorological processes and anomalous catch of the Asian pink salmon.

REFERENCES

[1] Bolotin E.I., Tsitsiashvili G.Sh. Spatial-temporal prognostication in functioning of center of mite encephalitis infection. *Bulletin of FEBRAS*, 2003. № 1, pp. 5–19. (In Russian).

[2] Shuntov V.P. New data on marine stage of life of Asian pink salmon. *Bulletin of TINRO*, 1994. Vol. 116, pp. 3–41. (In Russian).

[3] Summary on salmon fishery catch (Russian Federation). *Fish industry*, 1995. №. 6. 31 p. (In Russian).

[4] Summary on salmon fishery catch (Russian Federation). *Fish industry*, 1998. № 6. 28 p. (In Russian).

[5] Summary on salmon fishery catch (Russian Federation). *Fish industry*, 2000. № 6. 27 p. (In Russian).

[6] Tsitsiashvili G.Sh., Shatilina T.A., Kulik V.V., Nikitina M.A., Golycheva I.V. Modification of method of interval mathematics applicable to the forecast of extreme ice cover in the Okhotsk Sea. *Bulletin of FEBRAS*, 2002. № 4. pp. 111–118. (In Russian).

[7] Yanovskaya N.V., Sergeeva N.N., Bogdan E.A. *et al. Catch of Pacific salmon*, 1900–1986. Moscow: VNIRO, 1989. 213 p. (In Russian).

Translated by Vera Kochetova (Far Eastern State Technical University).

Chapter 2

APPLICATION OF EXPERIENCE METHOD OF THE RECOGNITION BY INTERVAL FOR MAKING PROGNOSIS ON THE TATAR STRAIT (JAPAN SEA) ICE-COVER EXTREMITY

T.A. Shatilina[1], G.Sh. Tsitsiashvili[2],**and T.V. Radchenkova***
* Pacific Research Fisheries Center, Vladivostok, Russia
** Institute of Applied Mathematics, FEB RAS, Vladivostok, Russia

ABSTRACT

The peculiarities of atmospheric processes in the middle troposphere that stimulate formation of anomalous ice cover in the Tatar Strait were explored. A clear distinction in AT500 fields was discovered above the central sector of the second natural synoptic region in the Northern Hemisphere under the small and large ice cover of the aquatic area of the northern part of the Japan sea. For the experimental prognosis of extreme ice cover in the Tatar Strait, the method of interval recognition was used. Analysis of the coefficients of recognition according to the data of above-earth pressure and H500 magnitudes for maximum and minimum ice cover in Tatar Strait exhibited that summer (June, July) and autumn (September, October) months possess the most distinguished recognition values that is in accordance with the physical explanation of reasons that cause formation of ice cover in this region.

INTRODUCTION

The Tatar Strait ice (Japan Sea) is one of the chains in the Far East climatal system. Variability of the ice cover is known to be dependent on the entire row of the factors: [1], [2], [3]. Despite voluminous data reports on the Japan Sea ice-cover, the reasons for its inter-

---

[1] shatilina@tinro.ru.
[2] guram@iam.dvo.ru.

annual and multi-year variability haven't been fully clear until now. In practice, such methodsas the quantitative analysis of anomalous ice-cover in Tatar Strait (as well as in the other Far Easten sees) aren't developed. New approaches to analysis and prognosis of ice-cover extremity for the Okhotsk Sea were offered and published by Tsitsiashvili G. Sh. et al. [4]. Some advantage of the interval recognition method for prognosis of extreme natural phenomena were demonstrated on the example of critical gorbusha catches in the Far East fishery regions [6].

The present work is aimed at verification of physical hypotheses of extreme ice-cover formation mechanisms in the Tatar Strait and possibility to prognosticate it with the help of interval recognition method.

1. DATA AND METHODS OF EXPLORATION

Data on the ice cover square (%) in the Tatar Strait (Japan Sea) that were published in Kryndin's work [1] during 1928-1960 and data received from the Far East Regional Center of admission and processing data (Khabarovsk) for the 1961-1995 period were used. Recent years (1996-2001) data were obtained from the Internet *http://www.natice.noaa.gov/*.

In our work we processed the data on the above-earth pressure during the 1948-2003 placed on the Internet (*http://dss.ucar.edu/data*) and in the archives of monthly average data on geopotential H_{500} that were recorded in CD-ROM "NCEP/NCAR Reanalysis Monthly Mean CD-ROM 1948-1998".

For the experimental method of extreme ice cover prognosis in the Tatar Strait, there was applied the method of interval recognition described in the works of Tsitsiashvili G. Sh. et al .2002 [4], Svyatukha et al 2003 [6]. The method was described in the first paragraph of this monograph.

2. PECULIARITIES OF THE ATMOSPHERIC PROCESSES THAT PROVIDE FORMATION OF ANOMALOUS ICE COVER IN THE TATAR STRAIT

Multi-year variability of ice cover area in January, February, March over the Tatar Strait is presented in the Figure 1.

In January the years of extremely large ice cover, namely, 1951, 1953, 1954, 1956, 1961, 1971, 1985 are displayed, so are 1963, 1974, 1981, 1991, 1993 but on the contrary, with low ice cover. In February the next years were assessed as ones with extremely large ice cover: 1951, 1953, 1954, 1960, 1979, 1985, but the extremely low ice cover was observed in 1957, 1963, 1974, 1981, 1991, 1993 (actually the same years of the January span). The same years with abnormal large ice cover area, like in January and in February, were identified in March.

In order to stress the patterns of extreme conditions icy formation in the Tatar Strait during 1948-2001, the atmospheric macro-process in the middle troposphere above the central region of the second natural synoptic area in Northern Hemisphere (30°-70° N, 120°-160°E) were researched. This analysis provided distinctive features of H_{500} fields demonstration under the low or large ice cover of aquatic area in the norther n part of the Japan Sea.

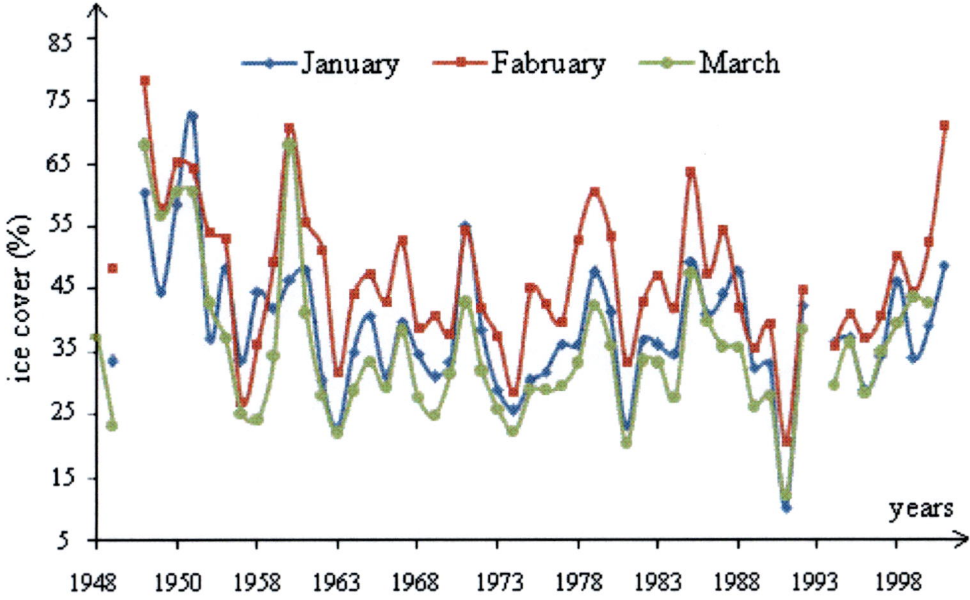

Figure 1. Inter-annual variability of ice cover (%) in the Tatar Strait during January, February, March of 1948-2001.

So, in January of 1954 when ice cover of the Tatar Strait above the Japan Sea (Northern part) was extremely large, there was observed well developed the Okhotsk Sea's middle tropospheric cyclone that conditioned formation of a cold core above the Japan Sea. In January of 1991, when ice cover was extremely low, this center of the Okhotsk Sea's cyclone was displaced into the Okhotsk Sea (Southern part) and the warm core was located over the Japan Sea.

Anomalous atmospheric processes that were watched in January of 1991, made conditions for extremely low icy cover in the Okhotsk Sea [5]. Figure 2 A, B exhibit difference in anomalies run according to latitudes on 140°E (it crosses the Tatar Strait) in January for the extreme years of the Tatar Strait ice cover. It is evident, that in 1963 and in 1991 (when ice cover was minimum) anomaly of H_{500} geopotential achieved 15 ∂am over the Tatar Strait, but in 1954, 1979 (when ice cover was maximum) it was minus 10 ∂am. The centers of core in positive and negative anomalies of H_{500} geopotential are shown, to be localized above the Tatar Strait (Northern part). Also, these atmospheric processes that were observed in those January years above the second natural synoptic region (NSR) were anomalous, although at the background of these large-scale anomalies the local extremes were fixed.

So, fulfilled analysis disclosed that the reason of extreme ice cover in the Tatar Strait are abnormal atmospheric processes running above the center region of the second NSR under which the cores of basic field anomalies are situated over the Japan Sea (Northern part). The cleared up difference in H_{500} fields during extreme ice cover years provided application of the method of interval recognition [4], [5], [6] for confirmation of atmospheric processes role to prognosticate experimentally the anomalous ice cover area in the Tatar Strait.

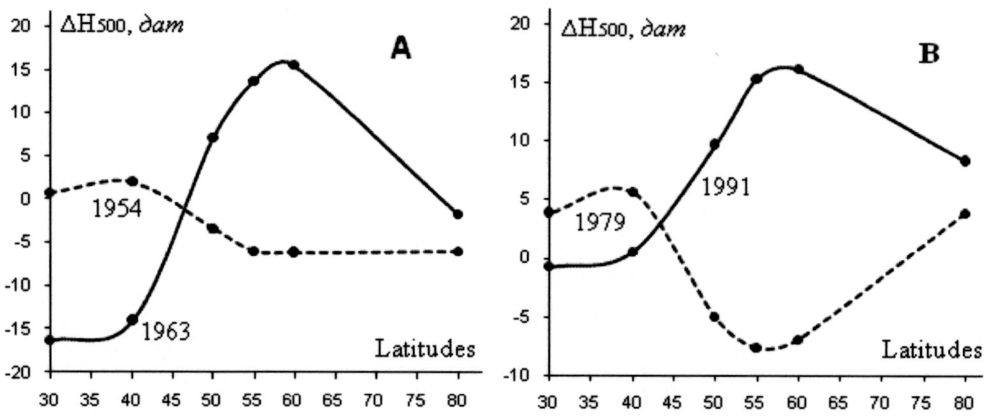

Figure 2. Latitudinal variability of anomalies in H500 geopotential along 140°E in January.

3. RESULTS OF CALCULATIONS

The central sector of NSR was divided into three climatal sub-zones, namely, northern (60°-65°N, 120°-160°E), temperate (45°-55°N, 120°-160°E) and southern (30°-40°N, 120°-160°E). Magnitudes of recognition coefficient (n) were determined for the points with H500 values, that were there before the extreme years of ice cover in the Tatar Strait. Estimation was conducted for all the months of preceding years. Analysis of these calculations demonstrated that every region is distinguished by its coefficient figures, i.e. it has distinctive informational features. The most information processing data (the highest number of 100% recognitions) to be reliable there, were the southern and temporal region. So, unification of these regions into one ensued in the highest number of recognitions, namely, 100% (Table 1).

Analysis of coefficients of recognition according to H500 values for maximum and minimum ice cover in the Tatar Strait discovered that the most (100%) recognition was in summer (June, July) and in autumn (September, October) months. This result corresponds to exploration of Stolyarova [3], whose opinion was, that the inter-annual fluctuations of water warmth and the Tatar Strait ice cover are influenced by advection constituents to a considerable extent. Therefore she tried to assess the influence of transfer of water and warmth acting on ice cover. Joint analysis of ice cover magnitudes in the Tatar Strait (average for January – March) with transfer of water and warmth along the transaction from Cape Zolotoy to Cape Slepikovsky showed distinctive connection being present between transfer of water and warmth and following then ice cover. According to our hypothesis the atmospheric circulation conditioned intensity of this transfer. Beyond this, Stolyarova expounded ice cover in the Tatar Strait to be identified at the great extent by might of water and warmth streams in the southern half - system of currents. Our computation disclosed that southern region possesses great informational parameters for the recognition of extreme ice cover (Table 1) that proves the physical hypothesis of ice cover formation in the Tatar Strait. It should be noted, that observations in systematic mode were absent along the hydrological transaction, therefore as the fundamental data for prognosis, there can be used regular information about atmospheric pressure which goes through the websites.

Table 1. Coefficients of interval recognition for the extreme ice cover (s, %) in the Tatar Strait according to H_{500} geopotential data during previous years

Month	January		February		March	
	$s\,\%>48$	$s\,\%<26$	$s\,\%>60$	$s\,\%<35$	$s\,\%>47$	$s\,\%<24$
January	0.7	0.71	0.54	0.67	0.75	0.71
February	0.78	0.55	0.86	0.6	0.86	0.55
March	0.58	0.45	0.75	0.4	0.75	0.45
April	0.7	0.83	0.75	0.86	0.86	0.83
May	0.78	1	1	0.86	1	1
June	1	0.83	1	0.86	1	0.83
July	1	1	1	1	1	1
August	0.78	1	0.86	1	1	1
September	1	1	1	1	1	1
October	1	1	1	1	1	1
November	0.87	0.71	0.75	0.67	0.86	0.71
December	0.64	0.83	0.86	0.86	0.6	0.83

The reliable information for extreme ice cover prognosis in the Tatar Strait that are used are based on the above-earth pressure over the same regions that were explored for the H_{500} fields (30°-55°N, 120°-160°E). The results of calculation are present in Table 2.

Table 2. Coefficients of interval recognition for the extreme ice cover (s, %) in the Tatar Strait according to the data of above-earth pressure during previous years

Month	January		February		March	
	$s\,\%>48$	$s\,\%<26$	$s\,\%>60$	$s\,\%<35$	$s\,\%>47$	$s\,\%<24$
January	1	0.83	0.75	0.67	0.86	0.83
February	0.58	0.36	0.86	0.37	0.86	0.36
March	1	0.71	0.86	0.43	0.75	0.71
April	1	0.71	0.43	0.55	0.67	0.71
May	1	1	1	0.86	1	1
June	0.87	0.83	0.75	0.86	0.86	0.83
July	1	1	1	1	1	1
August	0.78	0.71	0.86	0.67	0.86	0.71
September	1	1	1	1	1	1
October	0.87	1	0.86	1	1	1
November	1	0.71	1	0.5	1	0.71
December	0.7	1	1	0.86	1	1

The data from Table 2 testify to a greater information rate (presence of great number of units) of the above-earth pressure field in July, September, October that tallies with H_{500} geopotential data. Beyond this in contrast to H_{500} field, there was observed growth of the information rate of the above-earth pressure in November and December. The latter is concordant to physical concepts for the significant role of atmospheric processes in the ice-cover formation before winter in the Far Eastern Seas [1].

Table 3. Coefficients of the interval recognition according to H_{500} magnitudes in seasons

Seasons	January		February		March	
	s % >48	*s* % <26	*s* % >60	*s* % <35	*s* % >47	*s* % <24
Feb.-Mar.	0.88	0.71	1	0.75	1	0.71
Apr.-May	1	1	1	1	1	1
Jun.-Jul.	1	1	1	1	1	1
Aug.-Sep.	1	1	1	1	1	1
Oct.-Nov.	1	1	1	1	1	1

Table 4. Coefficients of interval recognition for non-extreme ice cover area (s, %) according to H500 field data in the Tatar Strait

Month	January		February		March	
	s % >44	*s* % <32	*s* % >53	*s* % <38	*s* % >40	*s* % <27
January	0.65	0.67	0.71	0.63	0.73	0.62
February	0.72	0.63	0.55	0.43	0.73	0.45
March	0.5	0.39	0.75	0.4	0.79	0.4
April	0.81	0.86	0.8	0.52	0.65	0.77
May	0.81	0.75	0.63	0.6	0.58	0.91
June	1	0.63	0.8	0.63	0.79	0.63
July	0.62	0.71	0.67	0.8	0.92	0.77
August	0.65	0.8	0.63	0.75	0.73	0.83
September	0.76	0.8	0.92	0.55	0.85	0.67
October	0.68	0.63	0.67	0.71	0.73	0.83
November	0.76	0.6	0.63	0.71	0.85	0.71
December	0.68	0.6	0.63	0.48	0.73	0.77

Especially high coefficients of recognition (100% in practice) were determined, when applying a method of interval recognition not only for several months, but according to the natural synoptic seasons as a whole, that were established comprising the character and features of the warm-exchanging process between the ocean and the continent of Asia. For the Far East the year was divided into six seasons, namely, the first half of winter (December – January), second half (February – March), spring (April – May), summer (June – July), autumn (August – September) and before winter (October – November).

The same high magnitude coefficients were obtained for the field of above-earth pressure on seasons. Such high coefficients of recognition could prompt supposition about inaccuracy of the method. Due to this method being directed to recognition of extreme events, the coefficients of interval recognition must be reduced under approximation of the main feature to the average magnitude. Actually, including into calculation the years without extreme ice cover such coefficient magnitudes "n" significantly reduced (Table 4). Based on this method it could be supposed possible to work for the extreme natural phenomena recognition.

Besides, we chose the most informative (based on physical hypothesis) region, where high coefficents of recognition proved the right choice of region. Also, calculation data from the other region (Aleutian depression area) evidenced the same (Table 5).

Table 5. Coefficients of interval recognition for the extreme ice cover area (S, %) in the Aleutian depression region according to H500 field data in preceding months

Month	January		February		March	
	s % >48	s % <26	s % >60	s % <35	s % >47	s % <24
January	0.19	0.15	0.24	0.15	0.29	0.15
February	0.17	0.28	0.15	0.24	0.16	0.28
March	0.23	0.5	0.27	0.5	0.22	0.5
April	0.54	0.11	0.29	0.13	0.29	0.11
May	0.27	0.22	0.27	0.25	0.32	0.22
June	0.27	0.13	0.24	0.16	0.22	0.13
July	0.2	0.31	0.32	0.32	0.2	0.31
August	0.33	0.5	0.3	0.27	0.32	0.5
September	0.32	0.21	0.21	0.24	0.25	0.21
October	0.3	0.18	0.32	0.2	0.3	0.18
November	0.17	0.16	0.15	0.19	0.15	0.16
December	0.19	0.33	0.38	0.4	0.33	0.33

To compare the results obtained by the method of interval recognition with the most often used methods we counted the coefficient of linear correlation between ice cover and the same data on pressure in congruous regions.

The coefficients of linear correlation are very low and have no prognostic meaning. Though, there should be stressed that the majority of coefficients more than 0.20, was observed in summer, corresponding to our computation and being evidence of the physical nature patterns reality that characterized ice cover in the Tatar Strait including the anomalous ones. Probably, one of the principal factors that make assists in the ice cover formation over the Japan Sea is the monsoon circulation, what in its turn provides transfer of the Tsusima current warm water into the Tatar Strait basin. The extremity of ice cover regimen is being formed under the abrupt alteration in the monsoon circulation regimen.

CONCLUSION

So, the analysis of atmospheric processes that are preceding anomalous patterns of ice cover in the Tatar Strait disclosed their difference. Application of the experience method of interval recognition for prognosis of extreme ice cover in the Tatar Strait demonstrated its great opportunity compared to the method of linear regression. Nowadays, evidently, the statistic prognosis with high reliability is available for three types of ice cover, namely, maximum, average, minimum. Prognosis could be predicted a few months in advance. In the future the testing method should be fulfilled on independent data.

REFERENCES

[1]   Kryndin A.N. Seasonal and inter-annual changes in the ice coverage and ice edge position in the Far Eastern Seas in connection with atmospheric circulation features//. *GOIN*, 1964. Vol. 781, p. 5-83. (In Russian).

[2]   Plotnikov V.V. Space-time relation of ice conditions in the Far Eastern Seas//. *Meteorology and Hydrology*, 1997. № 3, p.71-77. (In Russian).Stolyarova G.A. Influence advection to water masses of ice-conditions in the Strait of Tatar//. Proceedings of *FERHRI*, 1974. Vol. 55, p. 45 – 50. (In Russian).

[3]   Tsitsiashvili G.Sh., Shatilina T.A., Kulik V.V., Nikitina M.A., Golycheva I.V. The modification of the method of interval mathematics applicable to the forecast of extreme ice cover in the Okhotsk Sea//. Bulletin of *FEBRAS*, 2002. №. 4, pp. 111-118. (In Russian).

[4]   Shatilina T.A., Nikitin A.A., Muktepavel L.S. Features of atmospheric circulation at abnormal oceanological conditions in Japan, Okhotsk and adjacent Pacific// Bulletin of *TINRO*, 2002. Vol. 130, p. 79 – 94. (In Russian).

[5]   Svyatukha V.A., Tsitsiashvili G.Sh., Shatilina T.A., Goryainov A.A. Intra- annual bundles of climatic parameters//. *Pacific Oceanography*, 2003. Vol. 1, № 2, pp. 144-148.

Translated by Vera Kochetova (Far Eastern State Technical University).

Chapter 3

FACTOR TEMPORAL PROGNOSIS OF CRITICAL LEVELS OF HUMAN INFECTION RATE

E.I.Bolotin[1], G.Sh.Tsitsiashvili[2],** S.Yu.Fedorova* and T.V.Radchenkova***
* Pacific Institute of Geography, FEB RAS, Vladivostok, Russia
** Institute of Applied Mathematics, FEB RAS, Vladivostok, Russia

At present, there is the extensive scientific literature concerning the prognostic subjects. The analysis of this literature has been made time and again [1, 8] and there is no need to return to these matters again. It is necessary to mention some studies related to the medical prognostics including those appearing recently [4, 5].

It should be emphasized that, in the present work, we are dealing with the temporal prognosis of the wide spectrum of nosoforms of anthroponosic and zoonotic nature. As noted earlier, one can identify two real approaches to the temporal prognosis of human infection rate: extrapolation and factor prognosis. In this case, we have noted when using the tick-borne encephalitis model (see paper of these authors in the present volume) that quite a number of restrictions is after all characteristic of the extrapolation prognosis used extensively in different studies.

First, in case of the extrapolation prognosis, the averaged essentially and not existed in the nature parameters of the phenomenon under consideration are examined that reflect only a general tendency of developing this phenomenon with time. This trend is a smoothed artificial curve in which the most important for prognosis of catastrophic (critical) levels of the analyzed system characterizing the individual years or other time intervals were smoothed.

Secondly, the extrapolation prognosis results depend directly on the time series length and, what is more important, are determined by the type itself of multiyear movement of the phenomenon under study which could appreciably change its character at a different time.

Thirdly, the extrapolation prognosis results are of probabilistic nature and limited by confidence intervals. However, the range of these restrictions can be so wide that it,

---

[1] bolotin@tig.dvo.ru.
[2] guram@iam.dvo.ru

frequently, makes the results obtained so "vague" that it considerably restricts the possibilities of their objective interpretation.

Of the listed points, the second one is deserving of special attention that emphasizes a nonstationarity of time series of human infection rate and non-linearity of relations in the anthropoparasitogenic systems. It should be noted that the question of "non-linearity" is of prime importance and uninvestigated in the epidemiology and medical ecology while, at present, the interest in the nonlinear phenomena is very high and urgent [2].

A prognosis of "nonlinear systems" combined, at the same time, determinancy and stochasticity is an extremely intricate problem and, in this connection, the possibilities of prognosis of functioning such systems are quite restricted [6, and others]. To this, the opinion of the noted specialist in the field of statistics and prediction V.V.Nalimov should be added [7] who believes generally that the reliable scientific prognosis is only possible in its weakened variant (pattern-analysis) that comes to a thorough monitoring of events taking place in one or another analyzed ecological system.

The "factor" approach used by us in this work is based on the idea to predict the human infection rates that could be equal to or higher than a certain critical line specified by the researcher rather than some particular absolute indices. Such a statement of problem is principally important not only from the informal viewpoint because the extremely pressing problem of "nonlinearity" is removed but also methodically as the proposed way of its realization is of universal nature and can be easily reproduced both in the ecologo-epidemiological and other studies.

It is very significant that the set of factors used for a forecasting is "nominal" because the current level of our knowledge of causal relations in the anthropoparasitogenic systems is so far very limited in view of the extreme hierarchical complexity of these systems. In essence, realizing the factor prognosis with the use of some factors, we work in the dark resting only upon our experience and a priori conceptions of the possible relations in the systems under study. Strictly speaking, the factor prognosis should be preceded by the strong experimental ecological block of research studies in order to reveal true causal factors and mechanisms of their effect on the human infection rate or other epidemic indices. However, it is highly improbable that such a problem is solvable in the immediate future.

Nevertheless, setting a main task – development of principles and methods of the temporal factor forecasting – we use at this stage a certain set of factors which, on the one hand, are considered a priori causally related to the sickness rate while, on the other hand, are real representative long-term series of observations comparable with phenomena which are predicted by us.

More specifically, two groups of factors are used as effecting ones in this study: climatic (average annual temperature, absolute minimum temperature, length of frost-free period, days with the snow cover and maximum snow depth) and epidemic, i.e. long-term series of number of cases of different infection diseases. In addition, when predicting one or other infection, not all sets of other accompanying infections was used (though such a method is quite possible and practically realizable) but only those infections which are most similar in their character of long-term variations found earlier using a clustering [3].

The technique of the factor prognosis of critical levels of this or that infection rate was based on the following algorithm developed by us.

Initially, the empirical information of different nosoforms' rate and set of effecting factors characterizing one or another time interval and one or other territory (for example,

Primorsky Krai or only Vladivostok) were presented in the form of matrixes of initial data in which the rows are years while columns are indices of predicted human infection rates and effecting factors. Further, the years (rows) with critical levels of human infection rates and proper indices of effecting factors were identified. The identified years with critical levels of human infection rates form the ranges of particular values of effecting factors. However, the years with the human infection rates lower than identified critical level can also fall within these ranges. We will call such years "pseudocritical" for convenience. Let us denote the numbers of critical and "pseudocritical" years by "x" and "y" respectively. As a result, we obtain a relationship $p = x / (x + y)$ which can be interpreted as the probability to correctly identify (predict, recognize) the critical years according to the revealed intervals (ranges) of effecting factors.

As a whole, the developed technique of the temporal factor prognosis was based on the principles of pattern recognition. At the same time, the prognosis quality reached by the original algorithm was determined on the base of initial sampling by the number of uncritical (pseudocritical) years erroneously interpreted by our decision rule of recognition as critical ones.

The recognition-quality index (RQI) with the use of rectangular decision rule used in the study can be considered as the analog of multiple correlation between the predicted and effecting factors used in the mathematical statistics. Maximum recognition-quality index equal to unity implies the hundred-per-cent (100%) quality of recognition.

It is significant that the computation of this index is simpler than the estimation of the coefficient of multiple correlation made with the use of the method of principal components. In addition to a great body of computations determined by the number of effecting factors N, this method requires also large number of objects (i.e., years) n that should be much more than N.

In actual practice, in case of relatively small n (about 20), the value N is quite great (about 10 and more). The numerical experiments show that the attempt to increase n is provided, as a rule, by combination of heterogeneous samplings and, therefore, results in the recognition quality degradation rather than its improvement. In turn, the rise in critical level for the sign predicted and identification of the relatively homogenous sampling allow, as a rule, to increase the recognition quality index.

For the purpose of the prediction of exceeding the critical level of the human infection rate under analysis, the following modification of the algorithm was developed. At first, the intervals (ranges) for the revealed critical years for all effecting factors are constructed as per the algorithm described. Then, the number m of effecting factors each of which is attributed in a predicted year to a range (interval) corresponding to this factor is determined. If the value m coincides with total number of effecting factors N, therefore, as per the above rule, one can assume the excess of the critical level by human infection rate. However, if m is slightly lower than N one can expect an approaching the critical level. To verify the prognosis results, the critical and true levels of infection rates are compared for predicted years.

Consider some results obtained in the course of numerical experiments. Table 1 presents materials reflecting the recognition quality indices when predicting the critical levels of sickness rate for 24 leading infections in Primorsky Krai and Vladivostok with the use of 5 climatic factors of the weather station "Krasny Yar" located on the northern Primorsky Krai. In this case, the long-term dynamics of human infection rate was compared with the effecting factors both for the same year, i.e., without any time lag, and with a shift of 1 year. The time series for a period of 1975 to 2005 were considered.

As may be seen from materials presented in Table 1, the recognition quality of critical levels of human infection rate for Primorsky Krai as a whole and for Vladivostok varies from 31-40% (pertussis, gonorrhoea, salmonellosis) to 100% (ascariasis, enterobiasis, tick-borne rickettsiosis etc.). As a whole, one can consider that the recognition level for long series of infections predicted (more than 30 years) is not very high and its mean value for all infections reached 6.50 – 6.87. However, when using the shorter time series, the recognition quality increases considerably (table 2). So, the mean index of recognition for all infections reaches 0.85-0.92 and critical levels of about one-half infections are predicted with probability 100%.

Table 1. Recognition quality indices of critical levels of different infection rates using climatic factors of the weather station "Krasny Yar" for a period of 1975-2005

№	Nosoforms	Primorsky Krai		Vladivostok	
		Without lag	With lag	Without lag	With lag
1	Chickenpox	0.8	0.67	0.75	1.0
2	ARD	0.75	0.6	0.83	0.56
3	Influenza	0.8	0.56	0.8	0.83
4	Rubella	0.67	0.8	0.67	1.0
5	Infectious mononucleosis	0.86	0.6	0.8	0.8
6	Hepatitis A	0.83	0.56	0.83	0.56
7	Hepatitis B	0.75	0.55	0.67	0.86
8	Pertussis	0.31	0.71	0.4	1.0
9	Meningococcosis	0.64	0.8	0.83	0.83
10	Scarlet fever	0.55	0.36	0.45	0.45
11	Tuberculosis	0.83	0.71	0.86	0.6
12	Dysentery	0.54	0.67	0.5	0.56
13	Gonorrhoea	0.8	0.57	0.38	0.4
14	Syphilis	0.83	0.83	0.71	0.63
15	Ascariasis	1.0	1.0	1.0	1.0
16	Enterobiasis	0.63	1.0	0.71	1.0
17	Pediculosis	0.71	1.0	0.5	1.0
18	Scabies	0.45	0.63	0.56	0.63
19	HFRS	0.42	0.38	0.56	0.83
20	Tick-borne encephalitis	0.58	0.64	0.6	0.6
21	Pseudotuberculosis	0.6	0.55	0.55	0.67
22	Salmonellosis	0.4	0.46	0.47	0.41
23	Lime disease	0.56	0.83	0.45	0.63
24	Tick-borne rickettsiosis	0.5	1.0	0.71	1.0
	Sum of indices	15.81	16.48	15.59	16.46
	Mean value	6.59	6.87	6.5	6.86

Table 2. Recognition quality indices of critical levels of different infection rates using climatic factors of the weather station "Krasny Yar" for a period of 1990-2005

№	Nosoforms	Primorsky Krai		Vladivostok	
		Without lag	With lag	Without lag	With lag
1	Chickenpox	1.0	0.8	1.0	0.8
2	ARD	1.0	1.0	0.86	1.0
3	Influenza	0.83	1.0	0.83	1.0
4	Rubella	0.8	0.8	0.8	0.8
5	Infectious mononucleosis	1.0	1.0	1.0	1.0
6	Hepatitis A	0.71	0.8	0.57	1.0
7	Hepatitis B	1.0	0.83	0.71	1.0
8	Pertussis	0.63	0.8	0.83	1.0
9	Meningococcosis	1.0	0.8	0.56	1.0
10	Scarlet fever	0.63	0.67	0.63	0.67
11	Tuberculosis	1.0	1.0	1.0	1.0
12	Dysentery	1.0	0.8	1.0	1.0
13	Gonorrhoea	0.83	1.0	0.83	1.0
14	Syphilis	1.0	0.83	0.71	0.71
15	Ascariasis	0.83	1.0	0.83	1.0
16	Enterobiasis	0.83	1.0	0.71	0.8
17	Pediculosis	0.83	1.0	0.83	1.0
18	Scabies	0.63	0.71	0.63	0.71
19	HFRS	0.8	1.0	0.71	1.0
20	Tick-borne encephalitis	0.71	1.0	0.71	0.71
21	Pseudotuberculosis	1.0	1.0	0.83	1.0
22	Salmonellosis	1.0	1.0	0.83	1.0
23	Lime disease	0.71	1.0	0.63	0.83
24	Tick-borne rickettsiosis	0.63	1.0	1.0	1.0
	Sum of indices	20.4	21.84	19.0.4	22.03
	Mean value	0.85	0.91	0.79	0.92

The similar results concerning the recognition of critical levels of human infection rates were also obtained with the use of data of the weather station "Partizansk" located in the southern Primorye. In this case, a tendency to the recognition quality improvement was also revealed when decreasing the lengths of time series of human infection rates and effecting factors (Tables 3-4). At that, a considerable closeness in the results of predicting the critical levels of some infectious diseases with the use of meteorological data of two weather stations located at a considerable distance from each other and under different climatic conditions is explained by the similarity of a nature of long-term variations of these conditions recorded at different weather stations. Thus, it is believed that the quality of recognition of critical levels of predicted infectious diseases can be improved by way of selection of meteorological data of different weather stations.

Table 3. Recognition quality indices of critical levels of different infection rates using climatic factors of the weather station "Partizansk" for a period of 1975-2005

№	Nosoforms	Primorsky Krai		Vladivostok	
		Without lag	With lag	Without lag	With lag
1	Chickenpox	0.8	0.5	0.8	1.0
2	ARD	0.67	0.55	0.63	0.56
3	Influenza	0.8	0.83	0.8	1.0
4	Rubella	0.5	0.8	0.5	1.0
5	Infectious mononucleosis	0.67	0.75	0.8	1.0
6	Hepatitis A	0.71	0.38	0.71	0.38
7	Hepatitis B	0.55	0.86	0.67	0.67
8	Pertussis	0.71	0.83	0.5	1.0
9	Meningococcosis	0.41	0.67	0.83	0.5
10	Scarlet fever	0.5	0.33	0.5	0.42
11	Tuberculosis	1.0	0.71	0.67	0.75
12	Dysentery	0.54	0.43	0.6	0.83
13	Gonorrhoea	1.0	1.0	0.71	0.57
14	Syphilis	0.63	0.83	0.83	1.0
15	Ascariasis	1.0	0.5	0.45	0.83
16	Enterobiasis	1.0	0.83	1.0	1.0
17	Pediculosis	0.83	1.0	0.83	1.0
18	Scabies	0.5	0.71	1.0	1.0
19	HFRS	0.36	0.36	0.83	0.63
20	Tick-borne encephalitis	0.7	0.58	0.75	0.86
21	Pseudotuberculosis	0.86	0.75	0.86	0.55
22	Salmonellosis	0.67	0.55	0.58	0.44
23	Lime disease	0.71	0.71	0.5	0.83
24	Tick-borne rickettsiosis	0.71	0.83	0.83	0.83
	Sum of indices	16.83	16.29	16.68	18.65
	Mean value	0.7	0.68	0.7	0.78

As a whole, as a result of analysis of Tables 1-4, one can state the very important fact that each of 24 predicted nosologic forms has been recognized, at least, one time with the 100% probability.

Let us consider now the results of recognition quality of critical levels of different infectious diseases using the epidemic data as the effecting factors (Table 5).

Based on results presented in this table, one can identify several specific features in recognition of critical levels of particular infection rates belonging to different groups.

First, as in case of using the climatic factors, when the epidemic factors are applied, a tendency to an increase in the recognition level with shortening the series of infection rates is readily illustrated.

Table 4. Recognition quality indices of critical levels of different infection rates using climatic factors of the weather station "Partizansk" for a period of 1990-2005

№	Nosoforms	Primorsky Krai		Vladivostok	
		Without lag	With lag	Without lag	With lag
1	Chickenpox	1.0	0.44	0.83	0.44
2	ARD	1.0	0.67	0.86	0.83
3	Influenza	0.71	1.0	0.71	1.0
4	Rubella	0.8	0.8	0.8	0.8
5	Infectious mononucleosis	0.83	0.71	0.83	0.71
6	Hepatitis A	0.5	1.0	0.8	1.0
7	Hepatitis B	0.83	0.83	0.63	0.83
8	Pertussis	1.0	1.0	0.83	1.0
9	Meningococcosis	0.67	0.8	0.71	0.83
10	Scarlet fever	0.5	1.0	0.5	1.0
11	Tuberculosis	1.0	0.83	0.75	0.83
12	Dysentery	1.0	1.0	1.0	1.0
13	Gonorrhoea	1.0	1.0	1.0	1.0
14	Syphilis	0.71	0.83	0.83	1.0
15	Ascariasis	1.0	0.8	0.63	0.8
16	Enterobiasis	1.0	1.0	0.56	0.8
17	Pediculosis	1.0	1.0	1.0	1.0
18	Scabies	1.0	1.0	1.0	1.0
19	HFRS	0.57	0.57	0.83	0.63
20	Tick-borne encephalitis	0.83	0.71	0.83	0.83
21	Pseudotuberculosis	1.0	0.83	0.71	1.0
22	Salmonellosis	1.0	1.0	0.71	1.0
23	Lime disease	0.83	0.71	0.56	0.83
24	Tick-borne rickettsiosis	0.71	1.0	0.83	1.0
	Sum of indices	20.49	20.53	18.74	21.16
	Mean value	0.85	0.85	0.78	0.88

Table 5. Recognition quality indices of critical levels of different infection rates using epidemic factors for periods of 1975-2005 and 1990-2005, by groups of infections

Nos. of groups	Nosoforms	1975-2005		1990-2005	
		Without lag	With lag	Without lag	With lag
1	HFRS	0.31	0.42	0.8	0.67
	Tick-borne encephalitis	1.0	1.0	1.0	1.0
	Pseudotuberculosis	0.55	0.67	0.8	1.0
	Salmonellosis	0.5	0.5	1.0	1.0
	Lime disease	0.83	1.0	0.83	1.0
	Tick-borne rickettsiosis	1.0	0.71	1.0	0.71
	Sum of indices	4.19	4.3	5.43	5.38
	Mean value	0.7	0.72	0.91	0.9

Table 5. (Continued)

Nos. of groups	Nosoforms	1975-2005		1990-2005	
		Without lag	With lag	Without lag	With lag
2	Gonorrhoea	1.0	0.8	1.0	1.0
	Syphilis	1.0	1.0	1.0	1.0
	Ascariasis	1.0	1.0	1.0	0.8
	Enterobiasis	0.83	1.0	1.0	1.0
	Pediculosis	1.0	1.0	1.0	1.0
	Scabies	1.0	0.71	1.0	1.0
	ARD	1.0	0.86	1.0	1.0
	Sum of indices	6.83	6.37	7.0	6.8
	Mean value	0.98	0.91	1.0	0.97
3	Pertussis	0.71	0.45	1.0	1.0
	Meningococcosis	1.0	1.0	1.0	1.0
	Scarlet fever	1.0	0.63	0.56	1.0
	Tuberculosis	1.0	1.0	1.0	1.0
	Dysentery	0.71	0.44	1.0	1.0
	Influenza	0.57	0.8	1.0	1.0
	Sum of indices	4.99	4.32	5.56	6.0
	Mean value	0.83	0.72	0.93	1.0
4	Rubella	0.36	0.24	0.57	0.8
	Chickenpox	0.8	0.57	0.8	0.57
	Infectious mononucleosis	0.45	0.5	0.45	0.5
	Hepatitis A	1.0	0.83	0.83	0.8
	Hepatitis B	0.75	0.55	1.0	1.0
	Sum of indices	3.36	2.69	3.65	3.67
	Mean value	0.67	0.54	0.73	0.73

Table 6. Recognition quality indices of critical levels of HFRS and rubella rates with simultaneous use of climatic and epidemic factors for a period of 1990-2005

Factors	HFRS critical RQI level		Rubella critical RQI level	
Epidemic + climatic (Krasny Yar)	3.45	1.0	310	1.0
	4.0	1.0	330	1.0
	4.4	1.0	400	1.0
Epidemic+climatic (Partizansk)	3.45	0.83	310	1.0
	4.0	1.0	330	1.0
	4.4	1.0	400	1.0
Epidemic + climatic (both weather stations)	3.45	1.0	310	1.0
	4.0	1.0	330	1.0
	4.4	1.0	400	1.0

Secondly, the recognition of critical levels of infection rates for groups varies from 54-67% to 97-100% (4th and 2nd groups respectively).

Thirdly, when comparing different nosologic forms, a scatter of prognosis of critical levels is also very significant. For example, the tick-borne encephalitis is always diagnosed with the 100% probability in the first group of natural focal diseases while the level of HFRS recognition (diagnosis) is much lower and varies from 31 to 80 %. Such serious distinctions are also characteristic of the fourth group: for example, it is easily seen when comparing data of rubella and hepatitis A (Table 5).

In Table 6, materials reflecting the recognition quality of critical levels of rates of two characteristic: HFRS and rubella.

As might be seen from Table 6, three combinations of climatic and epidemic factors, particularly, epidemic factors and climatic parameters of the weather stations Krasny Yar and Partizansk (separately) and also their net effect were modeled. As a result, the hundred-per-cent quality of recognition was reached for all critical levels of infection rates with the exception of one case.

Therefore, one can state that the recognition increases essentially as a result of the net effect of factors of different nature. As a whole, the indices of recognition quality of critical levels of rates of different nosoforms can vary within very broad limits, from 30% and less to 100%. In this case, the major result of this portion of the work lies in the fact that a quite simple and convenient technique was developed whereby the hundred-per-cent recognition can be reached for all, without exception, infections used as prognostic nosologic forms by means of combining the effecting factors and modifying the lengths of infection rate series.

Let us consider now some results of the prediction itself of excess of critical level of infection rate analyzed using the climatic and epidemiologic factors (Tables 7-8). So in Table 7, a prediction of achievement or excess of the critical level of rates of different infections in Primorye in 2006 with the use of climatic factors and also verification of this prediction are given. For the particular example, the above climatic factors (5 voices-experts) of one of weather stations in the southern Primorye (weather station "Partizansk") with 16-year series (1990-2005) were used.

When analyzing this table, one can identify several most significant points of temporal forecasting with the use of climatic factors.

First, there was not a single case of "unanimous" vote out of twenty four when in case of prognosis ("for" cast) (see Table 7), all of five factors or, in contrast, no one of them would be recorded.

Secondly, only in seven cases of prediction (chickenpox, ARD, infectious mononucleosis, meningococcosis, tuberculosis, syphilis and tick-borne encephalitis), vote "for" was simultaneously made by four factors at the prognosis probability of 44 to 83 percent.

Thirdly, for seven above nosoforms, the prediction was correct in three cases (chickenpox, infectious mononucleosis and tuberculosis), it proved to be near real (ARD) and the prediction proved to be wrong in three cases.

Fourthly, for the most part, a vote "for" was made by two or three factors (of five) that makes difficult the objective interpretation of the prediction result obtained.

Table 7. Prediction of achievement of the critical levels of rates of different nosoforms in Primorye in 2006 using 5 climatic factors of weather station "Partizansk" over a period of 1990-2005

Nosoforms	Critical level (per 100 thousand)	Prediction (number of votes "for")	True level in 2006 (per 100 thousand)
Chickenpox	500	4	620
ARD	15300	4	13660
Influenza	1600	3	0.1
Rubella	310	3	34.8
Infectious mononucleosis	7.9	4	11.9
Hepatitis A	176	3	17.4
Hepatitis B	51	2	11.2
Pertussis	9.0	3	1.6
Meningococcosis	5.0	4	3.5
Scarlet fever	25	2	9.5
Tuberculosis	109	4	135
Dysentery	199	3	55.5
Gonorrhoea	149	3	46.9
Syphilis	200	4	86.5
Ascariasis	189	3	122
Enterobiasis	630	3	121
Pediculosis	300	3	109
Scabies	400	3	150
HFRS	4.0	3	2.3
Tick-borne encephalitis	6.5	4	3.4
Pseudotuberculosis	29	3	5.0
Salmonellosis	44	3	45.5
Lime disease	10.2	2	10.5
Tick-borne rickettsiosis	9.0	3	5.9

Therefore, the results of predicting the critical levels of diseases with the use of the above particular climatic factors and their values proved to be unsatisfactory in the majority of cases.

At the same time, the results of prediction using the epidemiological parameters present a principally different picture and are characterized by a quite high degree of accuracy against the background of prediction using climatic parameters. It must be emphasized that forecasting of the critical levels of one or other nosoform was performed in each case by only those concurrent infections which were most similar in their long-term variations revealed earlier by clustering and divided into 4 groups rather than by the whole set of other concurrent ones [3]. Thus, when predicting the particular nosoforms, five nosoforms were used as voices-experts in the first and third groups, six ones in the second group and four in the fourth.

It is seen from Table 8 that the "unanimous" vote was recorded in fourteen cases of twenty four, i.e. the number of votes "for" in all above cases was equal to zero. If the results of prediction for more three infections (syphilis, scabies, scarlet fever) are added when the number of votes "for" was equal to unity then one can recognize that prediction with the epidemiological factors was much more weighty than that with the use of "climate". The verification of the realized prediction for seventeen nosological forms showed that the prediction proved to be absolutely correct in thirteen cases while, in four cases (Lime disease, tuberculosis, Chickenpox and infectious mononucleosis), the prediction proved to be wrong because a true level of infection rate was slightly higher than critical one (see table 8).

Table 8. Prediction of achievement of the critical levels of rates of different nosoforms in Primorye in 2006 using epidemiological factors over a period of 1990-2005 by groups of infections

Nos. of groups	Nosoforms	Critical level (per 100 thousand)	Prediction (number of votes "for")	True level in 2006 (per 100 thousand)
1.	HFRS	4.4	2	2.3
	Tick-borne encephalitis	6.5	0	3.4
	Pseudotuberculosis	29	0	5.0
	Salmonellosis	44	2	45.5
	Lime disease	9.0	0	10.5
	Tick-borne rickettsiosis	10.0	0	5.9
2.	Gonorrhoea	149	2	46.9
	Syphilis	200	1	86.5
	Ascariasis	192	3	122
	Enterobiasis	630	2	121
	Pediculosis	300	2	109
	Scabies	250	1	150
	ARD	15700	2	13660
3.	Pertussis	9.0	0	1.6
	Meningococcosis	5.2	0	3.5
	Scarlet fever	28	1	9.5
	Tuberculosis	105	0	135
	Dysentery	200	0	55.5
	Influenza	1900	0	0.1
4.	Rubella	310	0	34.8
	Chickenpox	500	0	620
	Infectious mononucleosis	8.0	0	11.9
	Hepatitis A	176	0	17.4
	Hepatitis B	51	0	11.2

In conclusion, it should be noted that two presented examples of the temporal factor forecasting taken from several realized numerical experiments are of interest not only from the viewpoint of their conceptual essence. In our opinion, the main thing lies in the fact that the used approach offers vast, practically inexhaustible opportunities. In other words, one can achieve the satisfactory and even high quality of forecasting by means of unlimited examination or combining of forecasting variants, i.e. use of different effecting factors and their combinations, different lengths of time series and critical levels of infection rate levels and time lags etc. Thus, the opportunities and opened far-reaching prospects of the temporal factor forecasting of critical levels of infection rate of any etiology are demonstrated.

REFERENCES

[1]   Bolotin E.I. Functional organization of hot spots of zoonosic infections (by the example of the tick-borne encephalitis spots in the southern Far East of Russia). – Vladivostok : Dalnauka, 2002. (In Russian).

[2]   Bolotin E.I. On some debatable moments concerning the functional organization of the tick-borne encephalitis hot spots. *Parasitology*, 2006. 40 (6), pp. 547-555. (In Russian).

[3]   Bolotin E.I., Ananyev V.Yu. Space-time structure of human infection rate for the Russian Far East population: system approach. *Parasitology*, 2006. 40(4), pp. 371- 383. (In Russian).

[4]   Geltser B.I., Kukol L.V., Pupyshev A.V., Kolosov V.P. *Prognosis in pulmonology*. Vladivostok: Dalnauka, 2005. 182 p. (In Russian).

[5]   Grigoryev M.A., Grigoryev A.I. *Role of climatic factors in the short-term prognosis of tick-borne encephalitis rate* (by the example of Omskaya Oblast): Publ. house of Omsk State Pedagogical University, 2005. 196 p. (In Russian).

[6]   Kapitsa S.P., Kurdyumov S.P., Malinetsky G.G. *Synenergetics and predictions of future*. Moscow: Nauka, 1977. 285 p. (In Russian).

[7]   Nalimov V.V. Analysis of the ecological prediction fundamentals. Pattern-analysis as a weakened variant of prediction. In: *Man and biosphere. Ecological prediction*. Moscow: Mysle, 1983. Issue 8, pp. 31-47. (In Russian).

[8]   The workbook on forecasting. Moscow: Mysle, 1982. 430 p. (In Russian).

Translated by Victor Karpetc (Pacific Institute of Geography, FEB RAS).

In: Efficient Algorithms of Time Series Processing...   ISBN 978-1-60692-062-6
Editor: G. Sh. Tsitsiashvili   © 2009 Nova Science Publishers, Inc.

Chapter 4

SPACE-TIME PROGNOSIS OF TICK-BORNE ENCEPHALITIS FOCI FUNCTIONING

E.I.Bolotin[1,] and G.Sh. Tsitsiashvili[2,**]*
* Pacific Institute of Geography, FEB RAS, Vladivostok, Russia
** Institute of Applied Mathematics, FEB RAS, Vladivostok, Russia

The study of tick-borne encephalitis (TE) foci is a very urgent scientific-practical problem. It is related to the geography of the infection under study a natural habitat of which covers the vast forest and forest-steppe spaces of Russia and adjacent territories of Eurasia, high lethality, incapacitation, peculiarities of clinical implications and serious problems of population vaccination. It should be noted that, in connection with the TE problem significance, the international scientific forums are periodically organized the last of which was held in Vladivostok in autumn of 2007 and was devoted to the seventieth anniversary of studying this disease.

When studying TE that, in view of its specificity, is a good example of model infection, the prognostic aspects of the research study are of great importance. It should be noted that the most informative index of functioning of natural foci of TE is a specificity of human infection rate variations. This dynamic index is a function and certain result of impact of the whole set of abiotic, biotic and social factors. It should be noted that there exists extensive literature on the prognostic subjects including also a fair quantity of the fundamental summaries. According to different estimates, the total number of prognosis methods achieved at present 150 [8].

While on the subject of a prognosis of epidemic TE activity, it should be noted that this line of research is characterized by two aspects – spatial and temporal. Despite the fact that such division is quite conditional, in some instance, a similar differentiation is worthwhile and necessary.

---

[1] bolotin@tig.dvo.ru.
[2] guram@iam.dvo.ru.

We have identified the possibilities of the spatial epidemic prognosis using the TE model and proposed the algorithm of its realization. In this case, a realization of the territorial prognosis of epidemic situations (phenomenon A) in Primorsky Krai (in this case, prognosis of incidence of one or another clinical form of infection - a_0, a_1, a_2, a_3) was based on the methods of mathematical-cartographic modeling [2].

As an example (Tables 1, 2) of territorial prognosis of epidemic situation concerning the TE, let us consider two points of the Primorsky Krai territory sharply differing in the natural conditions: vicinities of Slavyanka settlement (south of Krai) and those of Vostok-2 settlement (north of Krai).

In the tables, the matrixes of coupling between the phenomenon under study and effecting factors characteristic of the mentioned localities of Primorsky Krai. A forecasting of probable epidemic situation in these localities was performed by summation of coefficients and showed that four predictable states are probable. In this case, a potential epidemic situation which is predicted for the northern Primorye is practically opposite to that in the southern Krai.

In essence, the used method of the territorial forecasting is similar to the judgment method the basis for which is formed by the principle of majority rule. A prognosis is performed by summation of these votes. If any states of the system obtain the same number of votes, a certain weight can be additionally given to them that is estimated by the value of measure of connection between the effecting factor and phenomenon [3].

Table 1. Prognosis of epidemic situation related to TE using the natural conditions of Slavyanka settlement (southern Primorsky Krai)

State of environmental factors	State of phenomenon (A)			
	a_0	a_1	a_2	a_3
B (high population)	1	0	0	0
C (low population)	1	0	0	0
D (high population)	1	0	0	0
E (low population)	1	0	0	0
F (more than 150 days)	1	0	0	0
G -(8-12 degrees)	1	0	0	0
I (0-200 m above sea level)	1	1	0	0
J (single specimens)	1	1	0	0
K (middle population)	0	1	1	1
L (2200-2500 degrees)	1	1	1	0
M (80-90 %)	1	1	0	0
Sum of prognostic points	10	5	2	1

Note. Here and in Table 2, population of Clethrionomus rufocanus (B), population of Haemaphysalis concinna (C), population of Apodemus agrarius (D), population of Ixodes persulcatus (E), length of a period woith positive temperatures (F), average January temperature (G), height above sea level (I), population of C.rutilus (J), population of A.peninsulae (K), sum of effective temperatures (L), relative air humidity (M).

Table 2. Prognosis of epidemic situation related to TE using the natural conditions of Vostok-2 settlement (northern Primorsky Krai)

State of environmental factors	State of phenomenon (A) a_0 a_1 a_2 a_3
B (single specimens)	0 1 1 1
C (high population)	0 1 1 1
D (single specimens)	0 1 1 1
E (high population)	0 1 1 1
F (up to 110 days)	0 1 1 1
G -(24-28 degrees)	0 1 1 1
I (more than 500 m above sea level)	0 0 1 1
J (high population)	0 0 1 1
K (high population)	0 0 1 1
L (up to 1600 degrees)	0 1 1 1
M (up to 70 %)	0 1 1 1
Sum of prognostic points	0 7 11 11

In the made examples of the spatial prognosis of the potential epidemic situation concerning the TE (Tables 1-2), 11 influential factors were used. Here, the fact of possible inclusion into the prognostic model of any factors the number and "nature" of which are determined by the representativeness of the used factors themselves and accuracy and reliability of achieved prognosis results is quite significant. The verification of the realized prognoses is made using the "reference" spatial points, i.e., those sections of the territory under study for which the actual condition of the phenomenon to be analyzed is known.

Let us consider now some resultant materials of the temporal prognosis of TE rate both for the Primorsky Krai territory on the whole and certain focal areas earlier identified by us [3].

It is important to emphasize that we are aware of the fact that the initial time series of the TE rate used by us are quite restricted and, evidently, not always reflect accurately the true epidemic situation, especially, in the past. Moreover, in this paper, the statistical materials on the TE rate after 2000 are not used for a number of reasons. In this connection, we believe just the methodical aspect of investigations will have the most priority, putting the development of methods on the first place while production of significant prognostic conclusions and evaluations only on the second one.

The analysis of the quite extensive scientific literature on the problem under consideration allows us to identify two real approaches to the temporal prognosis of the epidemic manifestation of natural foci of infections among which TE takes up a special position: extrapolation and factor temporal prognoses. We consider each of these methods (or groups of methods) separately.

The temporal extrapolation prognosis has been already applied when investigating the natural-focal infections [6] and it is realized, per se, on the principle of so called "black box" [11]. In other words, a realization of the temporal extrapolation prognosis is based on the idea that the used time series of infection rate (or other epidemic indices) represents a certain fixed result of the effect of totality of different influential factors in which a priori a certain prognostic information is contained [12].

As a material for the temporal extrapolation prognosis realized by us, the statistical data of the long-term dynamics of the TE infection rate in Primorsky Krai for different time periods were used. For the region as a whole, data for a period of 1940-1999 were used while for separate focal areas – data for a period of 1973 to 1999. The number of people being taken ill with TE within the focal areas in absolute and relative indices is presented in Table 3.

Table 3. The number of people being taken ill with TE for a period of 1973-1999 in 9 focal areas of Primorsky Krai in absolute and relative indices

№	Focal district	Absolute number of afflicted persons	On average per 100 thousand people
1.	Khasansky-Shkotovsky	282	1.7
2.	Nadezhdinsky-Ussuriisky	415	3.1
3.	Nakhodkinsky	112	2.1
4.	Partizansky-Lazovsky	131	4.9
5.	Spassky-Lesozavodsky	355	6.9
6.	Chuguyevsky	351	6.5
7.	Kavalerovsky-Dalnegorsky	182	6.7
8.	Dalnerechensky-Luchegorsky	71	5.3
9.	Central-Krasnoarmeisky-Pozharsky	176	15.6

The prognosis of the TE infection rate was performed using PC on the basis of an autoregression model - integrated moving average of Box-Jenkins (ARIMA), one of the most popular models used to process the time series of data and to construct prognoses [1]. From among a wide spectrum of computer programs of data processing including analysis of time series, here, the domestic statistical packet for processing of time series "Mezozavr" is used which is, in our opinion, is reliable and convenient in use.

Prior to discussing the results obtained, it should be noted that most considerable and serious contribution to the development and realization of temporal extrapolation prognoses concerning the TE infection was earlier made by R.L. Naumov with co-authors for eight large regions of the former USSR [6]. In doing so, it was emphasized that a lack of knowledge about the influence degree of different factors on the TE infection rate dynamics eliminates a possibility of the factor analysis. Within certain limits, this remark remains also correct at present. Such state of affairs is, in our opinion, related to the extraordinary complexity of studied natural foci of TE a functioning of which is determined by quite a wide spectrum of causal factors.

From the above reasoning, the realization of the temporal prognosis can be evidently made in two ways. First of them is extrapolation prognosis. It is based on the analysis of available time series of infection rate for the past years and their direct extrapolation, i.e., presents an analysis of one-dimensional time series.

The second approach to the temporal prognosis is, in essence, factor one and can be realized in several steps. In this case, strictly speaking, so called factors-indicators for which the levels of correlations with the infection rate dynamics are preliminarily identified rather than the true causal factors are used. The analysis of the correlation between dynamics of some natural-focal infections and different factors is presented in a number of well-known papers ([7], [9] and others). On the whole, the opportunities of the factor method of temporal prognosis being the multivariate analysis of time series as well as problems related to its application will be analyzed in detail below.

Coming directly to the temporal extrapolation prognosis, we emphasize some other important, in our opinion, initial conditions.

First, we, as also other authors, believe that, on the whole, the long-term variations of the TE infection rate level are of natural character while the most serious social factors (for example, different types of prophylaxis, population migrations etc.) make a corrective effect which does not disturb the natural tendency in the long-term course of infection rate.

Second, because it is evident that all medico-ecological processes are of cyclic nature ([10] and others) then an identification of the specificity of their variations and based on this prognosis depend, probably, on the observing time: the longer is empirical time series, the more objective, in principle, can be interpretation of its "behavior" and the more realistic should be prognosis.

Third, the use of extrapolation prognosis suggests certainly a procedure of verification of its results by means of "reference points", i.e., particular time intervals for which the actual state of the phenomenon under analysis is known. In this paper, a period of 1997-1999 is used as such time interval.

Figure 1 presents a number of graphs of smoothed time series of the TE infection rate in several focal districts of Primorsky Krai as well as the short-term prognosis of the infection rate (dotted lines) easily compared with actual data for a period of prognosis. The prognosis lines are surrounded with the upper and lower confidence limits.

First of all, it is necessary to note a very substantial moment residing in the fact that all presented graphs are very similar in character or tendency of long-term variations in the infection rate but differ greatly in the absolute indices. A similarity of the infection rate variation tendency manifests itself in the fact that in the 1980s the very low TE infection rate reaching the zero mark in some years was noted nearly everywhere while, in the 1990's, the abrupt increase in the infection rate with its stabilization towards the end of the decade came about. Generally, the similar type of the TE infection rate dynamics declared itself also in the West and East Siberia [4] that allows us to suppose the existence of some leading factor (or complex of factors) that determines the long-term variations of the TE infection rate on quite considerable territories. However, because a force of its effect (direct or mediated) on different territories is unequal, there are differences in the infection rate levels at the general similarity of the long-term dynamics nature.

The analysis of the realized temporal extrapolation prognosis results (Figure 1) allows us to assume a quite high degree of its representativeness. So, in seven focal districts (№ 1, 2, 4, 5, 6, 7, 8), the prognosis level proved to be very satisfactory, in one district (№ 9) satisfactory and, only in the Nakhodka focal district № 3), the prognosis differed substantially from real empiric data. Probably, the last case can be explained by a small sample of actual data and their more complex character of distribution.

As noted above, the accuracy of the extrapolation prognosis depends, most likely, on the length of the time series used. With the aim to check this hypothesis, we realized the prognosis of TE infection rate based on the time series of different lengths. In particular, the length of the basic time series reflecting the general infection rate for Krai changed with increments of 10 years, i.e. the time intervals beginning from 1940, 1950, 1960, 1970, 1980 were analyzed (Figure 2).

As became evident (Figure 3), the extrapolation prognosis based on the time interval from 1950 (a) proved to be the very accurate (close to 100%) while the prognosis for a period from 1940 (b) was characterized by a lesser accuracy. The practically opposite prognostic tendency as compared with actual data was found when analyzing the shortest time interval from 1980 (e), while the infection rate prognosis based on the time intervals from 1960 (d) and 1970 (c) occupied an intermediate position, besides, the shorter time interval provided the more accurate prognosis (Figure 3).

Therefore, one can ascertain that, within the framework of our investigations, the accuracy of the realized temporal extrapolation prognosis is, as a whole, determined by the length of the analyzed time series; i.e., the longer is the series, the more accurate is the prognosis. In this case, the essential corrective effect on the prognosis quality is made by the type of long-term infection rate variations.

The realized by us extrapolation prognosis of the epidemic manifestation of the TE foci in Primorsky Krai based on the analysis of one-dimensional time series of infection rate as well as the known studies for other territories allow us to suggest a certain perspective of this approach in the prognostic medico-ecological and epidemiological studies.

The analysis showed that even short time series which characterize the relatively homogenous in physico-geographical respect territories provide, as a rule, quite acceptable prognosis. In case of prognosis of the time series reflecting the infection rate on the great heterogeneous territories (Oblast, Krai etc.) and, especially, such contrasting ones as Primorsky Krai, the lengths of analyzed epidemic time series should be substantially increased. At the same time, it stands to reason that the extrapolation approach in the prognostication is characterized by a number of restrictions which determined our interest in more complex but, apparently, more objective factor approach in prognostication of the epidemic activity of the TE foci.

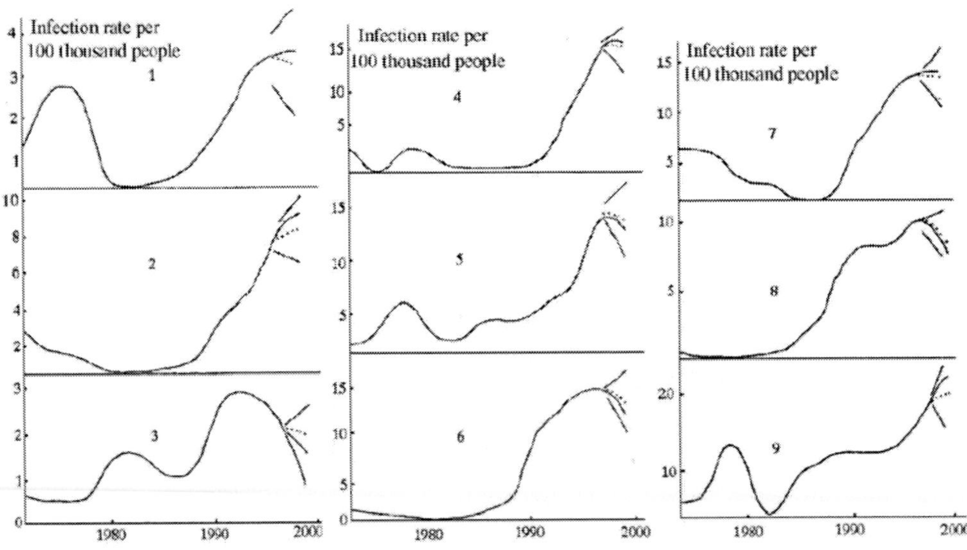

Figure 1. Graphs of smoothed time of the tick-borne encephalitis rate in 9 focal distincts of Primorsky Kray. Digits correspond to names of districts (see Table 3), dotted lines are extrapolation prognosis enclosed by the confidence limits.

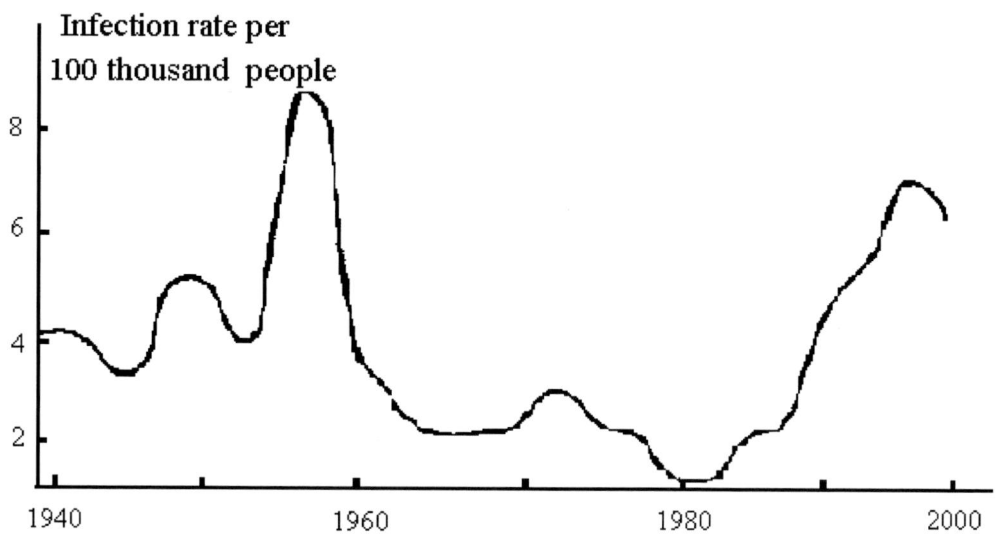

Figure 2. Graph of the smoothed time series of the tick-born infection rate divided into 5 intervals of different lengths.

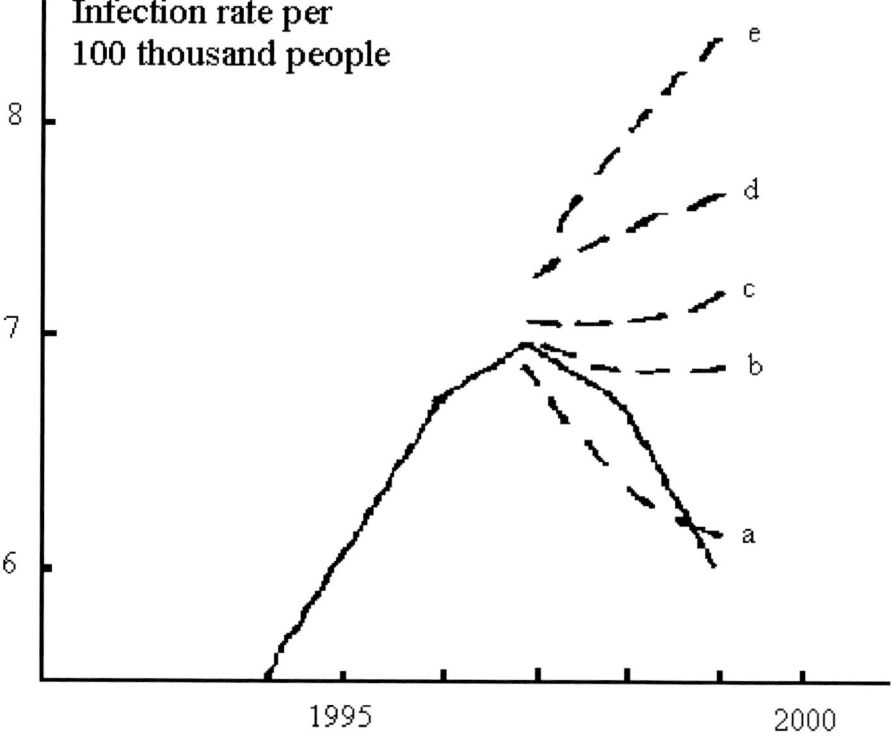

Figure 3. Extrapolation prognosis (dotted time) of the All-Krai tick-bome encephalitis rate based on ten time intervals of different lenghts (a-for a period from 1950, b-1940, c-1970, d-1960, e-1980).

However, as the specially carried out by us preliminary studies concerning the closeness and, most importantly, stability of ties between the long-term series of TE infection rate in Primorsky Krai and some expected influential factors showed, the identified levels of relations proved to be very unstable [3]. In other words, of each of the used influential factors, a considerable variability of the relation closeness occurred with the infection rate when changing the time series length or using different time intervals. Such a "pulsative correlation" between the infection rate and influential factors providing the very vague and inconclusive results restricts sharply the opportunities of the use of linear statistical models for realization of the factor temporal prognosis. Therefore, the idea has appeared to modify the statement of the prognosis problem. The main point of this idea implied that the levels of the infection rate which could exceed or be equal to some critical line specified by the researcher (expert) rather than particular absolute indices were predicted. In our opinion, such statement of problem is of methodical significance because the proposed way of its realization being universal could be easily reproduced both in ecologo-epidemiological and other studies. Of great importance is also the fact that the influential factors used for analysis can have any nature (abiotic, biotic, social, economic etc.) and their number is, in principle, unlimited.

The factor prognostic study of the epidemic manifestation of TE foci performed by us was divided into two stages. At the first stage, the prognostic retrospection of epidemiologic data was realized with determination of degree of accuracy of potential prognostic results when using different statistical samples. The object of the second stage of studies is to determine a degree of accuracy of factor prognosis of critical levels of TE infection rate realized in real time.

As a material for the factor temporal prognosis, the statistical data of the TE infection rate dynamics for 1973-1998 in a number of focal districts identified by us and regions of Primosky Krai served. At the first stage of prognosis, the influential factors being the long-term series of annual averages of eight meteorological factors obtained in three weather stations were used. These factors characterize the southern, middle and northern climatic-geographical zones of Primorsky Krai. The used meteorological factors include: average January temperature, absolute temperature minimum, average annual temperature, length of non-frosty period, average May temperature, average snow depth, number of days with snow cover and snow-temperature coefficient.

At the second stage of temporal factor prognosis, materials of the long-term dynamics of Ixodes persulcatus population recorded on three stationary routes within the suburban zone of Vladivostok were additionally used. Materials of the carrier population dynamics were only applied when predicting the infection rate in three southern focal regions of Primorsky Krai. In addition, in contrast to the first stage of studies when eight meteorological factors were used, the number of factors at the second stage, in connection with its specificity, was reduced to seven (factor "average annual temperature" was not used).

On the whole, the developed method of factor temporal prognosis was based on the image recognition principles. But the quality of prognosis realized by original algorithm was determined, using initial sample, by the number of non-critical years erroneously taken as critical ones by the final rule of recognition specified by us. At the first stage of investigations, the final rule of recognition of the year as critical was based on the condition of the hundred-per-cent falling of this year within the identified ranges of the effecting factors' values for critical years.

However, at the second stage of investigations, i.e. when predicting the critical events in real time, it turned out that such condition can play only auxiliary part in creation of more realistic rules of prognosis. This conclusion was made in the course of numerous numerical experiments carried out by us. They showed that, for construction of more realistic rules of prognostication, a mitigation of the "hundred-per-cent" condition is required. As a result, a new prognosis rule was established in which the number of identifying factors varied between 50 and 75 percent of all used ones depending upon the analyzed focal region, characterized by its specificity of the long-term dynamics of the TE infection rate. Critical infection rate was also chosen depending on the statistical sample with the number of critical years being more than one. It should be noted that the method itself of the factor temporal prognosis of the critical levels of the infection rate is not presented in this paper because it will be described in detail in our next work in these collected articles.

So, let us consider some results of the preliminary evaluation of the factor temporal prognosis of the TE infection rate.

The results of the first stage of prognosis for all identified focal territories of Primorsky Krai are given in Table 4. The maximum prognosis accuracy was found for the South-Primorsky focal region where it reached 81.8%. However, when differentiating this region into the focal districts, the prognosis accuracy increases and achieves maximum 100% index in Khasan-Shkotovsky and Nadezhdinsky-Ussuriisky focal districts and slightly lower (85.7%) in Nakhodkinsky focal district. A scatter of the prognosis accuracy for different territories of the Middle-Primorsky focal region is slightly larger: from 57.9% in Kavalerovsky-Dalnrgorsky to 100% in Partizansky-Lazovsky focal districts. In the remaining three focal districts of this focal region, the prognosis accuracy is the same and is on the average 71%.

Table 4. Accuracy of the retrospective prognosis (in %) of the TE infection rate for different focal territories of Primorsky Krai with the use of the complex of meteorological factors (The North-Primorsky focal region is, at the same time, Central-Krasnoarmeisky-Pozharsky focal district)

Focal regions and districts	Critical levels	Number of critical years	Number of "pseudocritical years"	Accuracy of prognosis
Focal regions:				
South-Primorsky	3	9	2	81.8
Middle-Primorsky	7	8	3	72.7
North-Primorsky	10	14	4	77.7
Focal districts:				
Khasansky-Shkotovsky	3	7	0	100
Nadezhdinsky-Ussuriisky	3	10	0	100
Nakhodkinsky	3	6	1	85.7
Partizansko-Lazovsky	7	6	0	100
Spassky-Lesozavodsky	7	11	5	68.8
Chuguevsky	7	8	3	72.7
Kavalerovsky-Dalnegorsky	7	11	8	57.9
Dalnerechensky-Luchegorsky	7	10	4	71.4

The results of the second stage of factor temporal prognosis are presented in Table 5. In it, a small fragment of the resulting materials reflecting the accuracy of factor temporal prognosis of critical levels of the TE infection rate realized in real time for different focal districts of Primorsky Krai is given. As the prognostic years, the time series for last 6-8 years were used depending on a character of the long-term dynamics of infection rate typical of one or another focal district. It should be noted that the presented data reflect only specially selected, most significant results concerning the prognosis accuracy of critical levels of the TE infection rate of the numerous variants of the realized temporal prognoses.

In the Table under analysis, the results of three variants of factor temporal prognosis are demonstrated. In the first case, four effecting factors were used: average January temperature, absolute temperature minimum, average May temperature and average snow depth. In the second case, to the above factors, three were added: length of frost-free period, number of days with snow cover, snow-temperature coefficient. And in the third case, the number of таежного клеща (I.persulcatus) was added to the factors listed. The number of critical levels for which the most significant results of the temporal prognosis in some focal districts were presented changed from one to three.

Based on the materials presented in Table 5, one can, on the whole, identify several most significant moments.

So, the accuracy of factor prognosis varied from 50 to 100%. At that, maximum accuracy level of a prognosis was observed for the southernmost territories of the region under study though some other geographical areas characterizing by the similar indices of prognosis (for example, Kavalerovsky-Dalnegorsky focal district located in the north-eastern Primorye) were identified.

The considerable effect on the prognosis level is made by the number of recognizing factors and their particular set. However, the increase in the number of used factors does not always result in the higher prognosis accuracy. This remarkable fact requires the further thorough investigation and refinement.

The obtained data concerning the relation between the prognosis accuracy and used critical levels of infection rate are also of interest. It was found out that in some cases this relation can be direct and in the others - indirect. This quite significant fact requires also a serious analysis and understanding.

It was revealed that the use of factual information of the long-term Ixodes perculcatus population dynamics had not improved the prognosis accuracy. It remained unchanged or even lowered. This fact confirms again our opinion [3] about the insignificant influence of the carrier population on human's suffering from tick-borne encephalitis and, therefore, on the intensity of natural foci of the infection under consideration.

If the results of the retrospective prognosis (Table 5) and prognosis in real time (table 4), i.e. results of the first and second stages of the factor temporal prognosis are compared, then, on the whole, one can note their considerable similarity. Some exception is the prognosis results for Kavalerovsky-Dalnegorsky focal district which can be a subject of the further, deeper analysis.

On the whole, the proposed method of the factor temporal prognosis of critical levels of infection rate allows us to solve the extremely nagging problem of "non-linearity", on the one hand, and is characterized by a number of fundamentally significant features.

Table 5. Accuracy of three variants of factor temporal prognosis (in %) of critical levels of TE infection rate for different focal districts of Primorsky Krai in real time

Focal district and critical level of the infection rate	4 factors	7 factors	7 factors and ticks
Khasansky-Shkotovsky			
3.2	100	100	83
Nadezhdinsky-Ussuriisky			
3.6	100	100	100
4.6	60	75	50
6.0	50	67	67
Nakhodkinsky			
3.5	63	63	50
5.5	75	75	75
6.0	75	50	50
Partizansky-Lazovsky			
5.0	100	100	
10.0	50	50	
Spassky-Lesozavodsky			
8.5	67	83	
10.0	83	83	
Chuguevsky			
7.5	83	67	
10.0	50	67	
Kavalerovsky-Dalnegorsky			
7.5	100	100	
10.0	86	100	
Dalnerechensky-Luchegorsky			
8.0	86	86	
14.0	71	57	
Central-Krasnoarmeisky-Pozharsky			
14.0	63	75	
26.0	88	75	

First, this method is quite easy for realization but, at the same time, universal, i.e., it is able to "work" with any information presented in the form of dynamic series.

Second, this method is "transparent" because all calculations performed by means of the program prepared can be quickly and clearly realized by hand.

Third, according to the classification J. Martino, known and authoritative expert-predictor, the proposed method can be considered as one of the highest steps of real prognosis based on the dynamic cause-and-effect models.

Fourth, the method of analyzing the critical levels of infection rate being, in our opinion, of theoretical and practical interest has good perspectives because it puts forth immediately a number of fundamental questions. Answering them in our further studies we will in perspective obtain a good method of the factor temporal prognosis.

REFERENCES

[1] Box D., Jenkins G. Analysis of time series. *Prognosis and control.* Vol. 1. Moscow: Mir, 1974. 406 p. (In Russian).

[2] Bolotin E.I. Peculiarities of the tick-borne encephalitis foci in the southern Far East. Vladivostok: Dalnauka, 1991. 96 p. (In Russian).

[3] Bolotin E.I. Functional organization of hot foci of zoonosic infections (by the example of the tick-borne encephalitis foci in the southern Far East of Russia). – Vladivostok: Dalnauka, 2002. 150 p. (In Russian).

[4] Zlobin V.I., Gorin O.Z. Tick-borne encephalitis. Novosibirsk: Nauka, 1996. 177 p. (In Russian)

[5] Martin J. Technological prognosis. Moscow: Progress, 1977. 592 p. (In Russian).

[6] Naumov R.L., Gutova V.P., Fonareva K.S. The coincidence degree of the long-term extrapolation prognosis with real tick-borne encephalitis rate. *Medical parasitology*, 1990. № 5. P.40-43. (In Russian).

[7] Neronov V.V., Malkhazova S.M. The analysis of relations between the zoonosic cutaneous leishmaniasis rate in the Murgab oasis and hydrometeorological factors. *Medical parasitology*, 1999. № 3. P.22-26. (In Russian).

[8] The workbook on forecasting. Moscow: Mysle, 1982, 430 p. (In Russian).

[9] Rothschild E.V., Kurolap S.A. Prognosis of the zoonosis foci activity. Moscow: Nauka, 1992. 184 p. (In Russian).

[10] Chizhevsky A.L. Earth echo of solar storms. Moscow: Mir, 1973. 349 p. (In Russian).

[11] Ashby U.R. Introduction in cybernetics. Moscow: Mir, 1959. 432 p. (In Russian).

[12] Yagodinsky V.N. Dynamics of epidemic process. Moscow: Medicina, 1977. 240 p. (In Russian).

Translated by Victor Karpetc (Pacific Institute of Geography, FEB RAS).

Chapter 5

SYSTEM APPROACH IN DEMOGRAPHIC INVESTIGATIONS

Z.I. Sidorkina[1], and G.Sh.Tsitsiashvili[2],***
* Pacific Institute of Geography, FEB RAS, Vladivostok, Russia
** Institute of Applied Mathematics, FEB RAS, Vladivostok, Russia

ABSTRACT

Under the conditions of unstable demographic indicators, when grounding administrative decisions, the most difficult thing is definition of their orientation. The offered analysis of network integrated schedules of passage of various generations through economic changes in a course of time has shown that there are branches in demographic system that are the most unstable to external pressure

At the beginning of economic reforms it was considered, and demographers agreed, that basic problems of the demographic future on the nearest 20-25 years are not so much in quantitative changes, as in qualitative factors. An abrupt drop in demographic indicators was not foreseen. A new conception of social policy, taking into account weak controllability of demographic processes, supposed minimization of consequences of crisis for family and society, though isolated researches gave warning about the ambiguous conduct of the demographic system [1].

During realization of economic reforms, demographic indicators began to change unexpectedly fast and brought absolutely unexpected consequences. It turned out that demography is to predict in its ordinary "time horizon" equal the length of one generation. All hypotheses of demographic changes somehow or other are built on the basis of past demographic events, discovered rules of dynamics and forecast demographic setting with one or another assumption. The Russian school of prognostication usually follows next consecution: the theoretical analysis of development of future events, creation dynamic

---

[1] zsidorkina@mail.primorye.ru.
2 guram@iam.dvo.ru.

programming model, the estimation of relative demographic coefficients and then on the basis of these estimations prognoses are built. For example, on the basis of study the dynamics of total number and tendencies of demographic changes in the region, it was offered the calculation of population of the coming generation, on the nearest 20-25 years, taking into account possible positive changes in a birth-rate and breaking of negative tendencies in a death rate. As per such calculations it is most likely we should expect depopulation in all subjects of the Far-Eastern federal district.

Under adverse conditions the population of the Far East by 2050 will be 2.17-2.5 million only (the same as the population of the Primorskiy area in 2000). The number of habitants in the Magadan area can became five times smaller, in Kamchatka, Sakhalin areas – only 2/3 less; in the Primorskiy area the population will shrink to 1.5 million. The results of this analysis are shown in monographs [3, 4]. Economic reforms, that started at the end of the eighties, effected social processes in such a way that demographic indicators in the ordinary range of one generation began to be estimated with difficulty by the traditional analysis. In this work an attempt was made to use a system approach to the estimation of demographic data on a regional level. The migration was ignored in this process. All analysis of dynamics is based on natural reproduction. Birth-rate and death-rate characteristics were selected for analysis as the most dynamic demographic indicators in the given unstable situation. They are reliable, have being analyzed for every year, and the migration has not had a noticeable impact on them. A system approach lets us avoid large errors.

In conformity with long duration tendencies, we made an attempt to approach the demographic system differentially. We estimated, in fact, not integral, but differential characteristics. It became clear at more detailed study of all data file, that it is necessary to work with disaggregation models, i.e. to take data for separated age groups and research the dynamics for each of them.

As shown in previous works of authors, the demography is characterized by irregular and impulsive changes [3, 5]. Under the conditions of la ong-term stable situation, parameters of the system almost do not change. But under the intensive influence of external factors, divergence of parameters are increasing and the system gets a new behavior, so called, unsteady branches. Even their minimal effect is enough, to have influence on the march of events (dynamics of process) [2]. Unsteady branches can radically change the nature of a stable system and result in huge and even tragic consequences.

On the short segments of a dynamic row, we have the opportunity to see unexpected distinctions, in our case this is distinctions in a birth-rate and death rate. These natural processes are changing over time in age groups as well; a combined diagram fixed it (Figure #1). The male death rate in all age groups much more than the female one. This index is especially obvious in 1959 in groups aged 50 and older. Men at age 55 had the same level of death rate, as women 65 years old, and in 1995 – as women 68 years old. The discrepancy between the male and female death rate have become even greater by 2000. The men aged 23 years die with the same frequency, as women at 50 years. But it is difficult to judge about the dynamics of the death rate in every age group using this diagram. A detailed elaboration of the system is required which could reduce uncertainty.

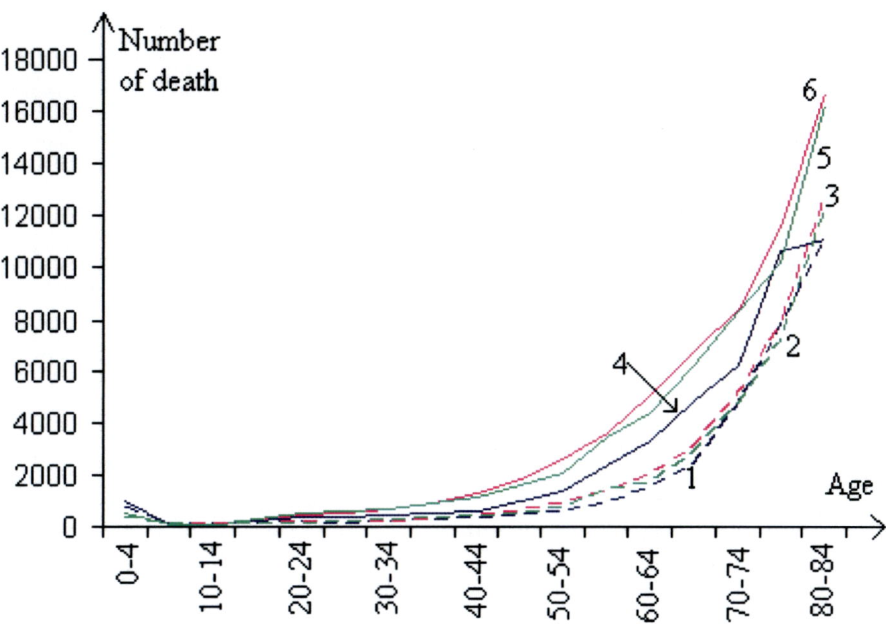

Figure 1. Male (4 – 1959, 5 – 2000, 6 -1995) and female (1 – 1959, 2 – 2000, 3 – 1995).

MORTALITY IN PRIMORSKIY KRAY DEPENDING ON AGE

It was accepted in demography to consider, that a death rate is high for children under 1 year, aged one to five years and after 60 years.

But such tendency changed after the economic crisis in the 1990s. Having appealed to the dynamics of total coefficients of birth-rate and death rate, we were studying internal behaviors of the demographic system by method of systems analysis, when the general index is broken up into fragments in a one-dimensional model: one coefficient is one age interval [4]. On comparing the same name coefficients in different age groups, we concentrated our attention on instability of coefficients within homogeneous groups. Such detailed elaboration let us raise reliability of analysis results and define, what age interval is subjected to the changes more and as they influence on other demographic behavior. It allows one to use a systems approach for population forecasting.

We built the diagrams of indexes of birth-rate and death rate for sixteen male and female age groups for the period 1959 - 2000. The various dimensions of scales in the diagrams let us see drastic changes of the dynamics that were not observed under traditional analysis. Using the network graphs, built up for a long period, it became possible to see those weak links which should be taken into account while creating the necessary prerequisites for population forecasting.

Upon the baby bust and increasing of the death rate in 1979-1995 the situation in most age groups began to be stabilized, and the diagrams show it.

The birth-rate went down and stabilized on quite a low level. With all this going on in most female groups of childbearing age, a growth trend in the birth rate is observed , except for females aged twenty to twenty-four, the most productive group, where the birth rate continues dropping (Figure 2). Whereas, in accordance with the coefficient of age-specific birth-rate, exactly this group stipulates the possibility of a population upsurge in the future, for example to 2015, due to birth of first, second and, possibly, subsequent children.

Increase of the death rate in the Primorskiy Kray has suspended for most ages after 1995, excepting young people aged 20-30 years (Figure 3).

The persons of this age were not able to overcome the negative influence of consequences of the economic crisis in the beginning of the 1990s. Since 1991 three age stages were replaced in the group of 20-24 years, but the coefficient of death rate did not only go down, but continues to rise. In the nearest years present children will pass into the age groups with a large death rate, and that is the reason why the index of the expected life interval for a population in from age 15 years will be a determining factor during realization of the social programs.

Strength of the death rate (probability to live so long to the next age group) turned out to be quite a complex value in population forecast calculations, in spite of the fact that this procedure is most developed. Traditionally, groups 20-24 and 25-30 years had the lowest death rate, but in the post-reformation period unexpected changes were observed, which are to be analyzed in order to avoid problems in defining a further tendency. Female mortality aged 55-59 is growing as well (Figure 4).

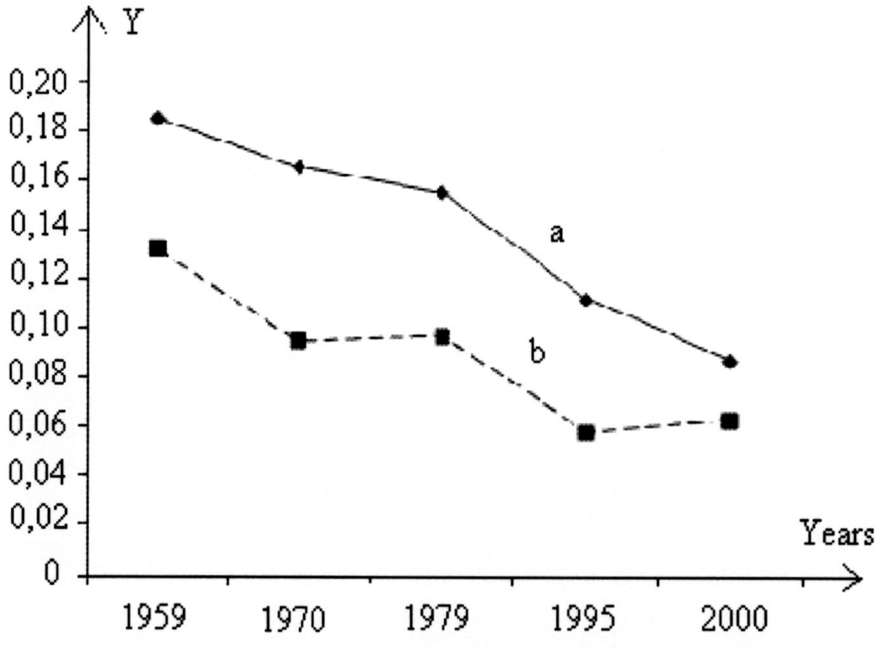

Figure 2. Dynamic of female age-specific birth rate in Primorskiy Kray: a – aged 20-24, b – 25-29, Y-line – birth rate coefficient on 1000 p.

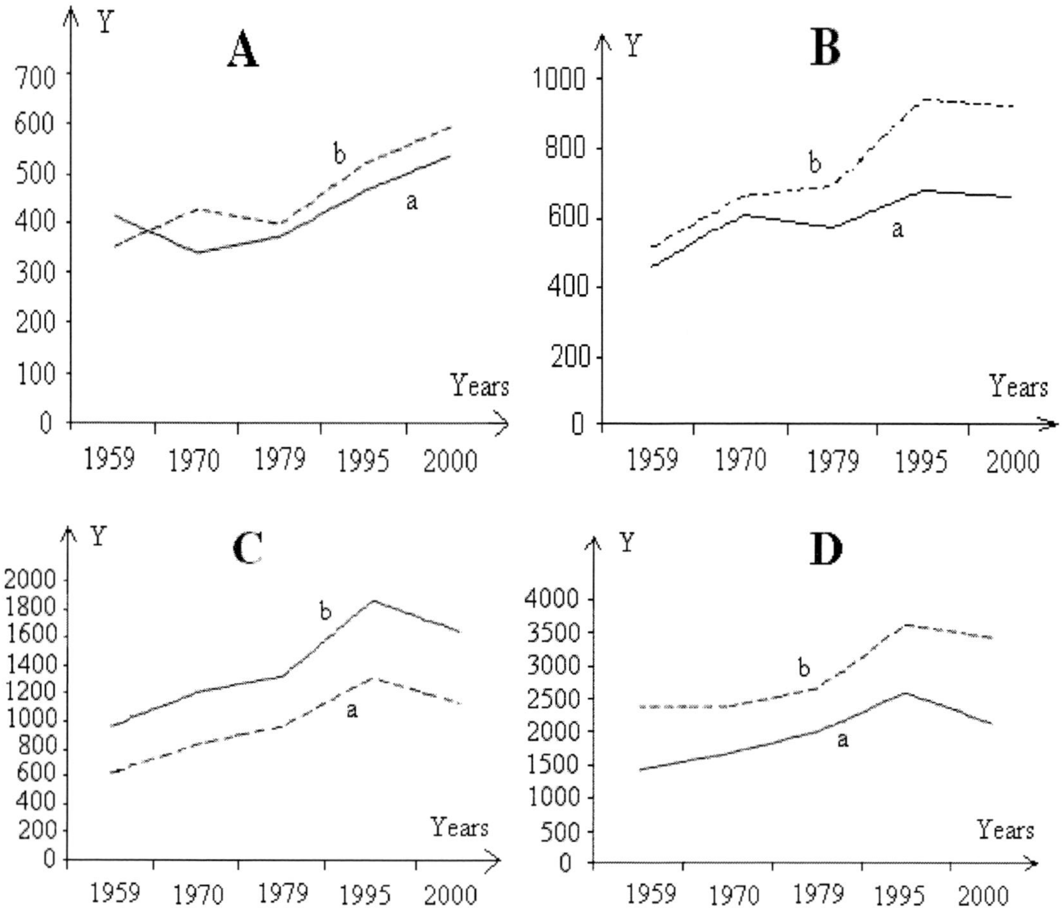

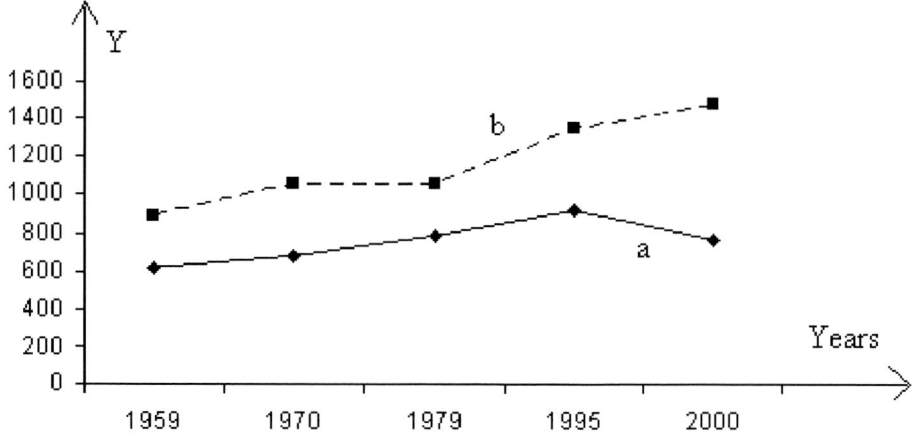

Figure 3. Male mortality in separate age groups in Primorskiy kray: Y-line – number of death per 100 000 persons, A: a – aged 20-24, b – 25-29; B: a – 30-34, b – 35-39; C: a – 40-44, b – 45-49; D: a – 50-54, b – 55-59.

Figure 4. Female mortality in Primorskiy kray: a – aged 50-54, b –55-59.

It is obvious, that changed tendencies of mortality indices for young people 20-30 years and women older than 55 years must become determinant at the prognosis of quantity of population and determination of priorities of the demographic programs.

Thus, our approach let us analyze the variants of development of demographic events more in detail in order to rightly take into account possible changes. Research results are evidence of the necessity of state support of the most vulnerable groups in the population and taking urgent measures to break the current demographic situation.

REFERENCES

[1]   Besstuzhev-Lada I.V. *Forecasting grounds of social innovation.* Moscow: Nauka, 1993. 240 p. (In Russian).

[2]   Kurlov A.B. System sociodynamics of present-day society (experiment with correction of deterministic ontology social society). Vestnik of Ufimskiy GATU (Scientific articles and reports. Humanitarian and Social sciences), 2000. № 2, pp. 23-32. (In Russian).

[3]   Avdeev Y.A. The perspectives of socio-demographic development of Primorskiy kray till 2025. Vladivostok: Dalnauka, 2004. 170 p. (In Russian).

[4]   Avdeev Y.A. *Population Problems of the Far East.* Edited by Avdeev Y.A.Vladivostok: Dalnauka, 2004 . 214 p. (In Russian).

[5]   Tstsiashvili G.Sh., Sidorkina Z.I. *The using of system analysis method in demographic forecast.* Preprint. Vladivostok: Dalnauka, 2005. 24 p. (In Russian).

Translated by Pavel Sidorkin (RoKcs/Vladivostok).

ISBN 978-1-60692-062-6
© 2009 Nova Science Publishers, Inc.

Chapter 6

THE DETERMINATION OF FIXITY FACTORS IN DYNAMIC RISE OF CITIES

***Z.I. Sidorkina**[1,*] **and G.Sh.Tsitsiashvili**[2,**]*
* Pacific Institute of Geography, FEB RAS, Vladivostok, Russia
** Institute of Applied Mathematics, FEB RAS, Vladivostok, Russia

ABSTRACT

The article is devoted to analysis, performed by means of an integrated operational schedule, of population changes in the cities of the Russian Far-East with various populations in time domain containing crucial moment of development. The graphic analysis with the different size of scales has let usdelimit aggregated dynamics on two clusters – the cities with decreasing and with a stable population.Two main factors that have had a positive influence on the dynamics of the development of cities were clearly recognized. They are – frontier location and the presence of the government enterprises, which are extremely important for the Russian state.

Territory of the Far East is the most urbanized among federal districts. The share of urban population is 80%. There are only two cities with a population over 500, 000, five towns belong to the group with population over 150,000, all other towns have a population of less than 100 000 habitants. However, the importance of the last group for the settlement pattern and, in the first turn for the settlement of town-forming significance, is much more than in more densely populated regions. The main objective of this work is to show changes in the structure and dynamics of the urban settlement pattern as a result of reformation and what its trends are. Population changes are the most generalized indicator of the social and economic potential of a city. That's why one condition of the solution intended task is offering a more simple method of assessment of the present and the nearest future of the cities.

This let us reach a quick enough conclusion about the state of population movement. The method, tried by us, is based on a pure econometric approach. It holds the studies of the

---

[1] zsidorkina@mail.primorye.ru.
[2] guram@iam.dvo.ru.

settlement population dynamics in a time lag, containing the critical moment of development, assuming pure visual classification: either stabilization in the neighborhood of some value or a lessening. In contrast to the ordinary, traditional approach, we were using a different dimension of scales in population dynamics graphs [1]. In that way we could notice essential changes which were not visible, using the same type scale.

Population number dynamics were analyzed in cities of the Far East for the period 1959-2002, on the whole. This led us to present by visual demonstration all situation complexity in cities population changes.

In the cities with population over 500,000 (Vladivostok and Khabarovsk), population dynamics have become negative by end of the 90s. Khabarovsk is losing its demographic potential slower than Vladivostok. These cities are more inviting for people because of better chances of being placed in a job, but due to high living expenses migrants, moving to the southern part of the region, have to settle in their suburbs, where rental business (habitation, land, trading etc.) has become widespread.

The market price of plots is determined by remoteness from central cities and main roads. So, the more distance the lower the price of the plot. The cost of plots with a house in Vladivostok divided into price zones: the first zone was directly in a city's limits, the second zone of prices – from the border of compact residential building up to 10-15 km off city's center (forest-park area). The cost 0.01 ga without any structures or with a tumbledown one, equal to the cost of an apartment in a tenement house. In a radius of 15-30 km from a center (third price zone ends) – the same plot is estimated at 2 times below, that in the second zone. In a radius of 30 - 45 km (fourth, with country type of the suburban settlements) the same plots cost ten times below, that in the first and second zones. Cost of plots with summer houses in a country place, situated along a railways 50 km away from a city and more is tens times less as well.

The cities Blagoveshchensk, Komsomol'sk-na Amur and Yuzhno-Sahalinsk (each one with population over 200 thousands) are equal in size and are stable enough in the development. However, Blagoveshchensk is more stable compared to Комsomolsk-na-Amure. The last one had declining population in the first half of the 90s. The population decline was stopped in 2000 as a result of a flow of investments into local industry and applications of new technologies that improved the production quality.

The population in the city of Yuzhno-Sakhalinsk is retaining stable on account of investments into oil-and-gas production. The population is concentrated in the northern part of Sakhalin Island. Petropavlovsk-Kamchatski city became the most vulnerable among other cities in this group. Its population declined considerably and now is the same as at the beginning of the 60s (Figure 1).

The cities of Ussuriysk and Nakhodka belong to the group with population (under 200 ths. people) and they had positive population dynamics in the 90s. They remain rather stable and have kept the population on the same level.

Under low density of large-scale cities (one large city on five small ones) and considerable remoteness of small cities from the centers of federal districts, small towns themselves became control centers, managing the social life of surrounding territories. The location of public enterprises in such small towns promoted their quick development (Zeya, Amursk, Arsen'yev, Dalnegotsk) [2] As a result of concentration of factories, oriented on federal or regional market, small towns turned out to be wittingly more dependent on the swing in production than large cities. In conditions of formation of a market system, disadvantages of small towns, resided them as a type of settlements (mono-profile functional organization) became more aggravated (Figure 2).

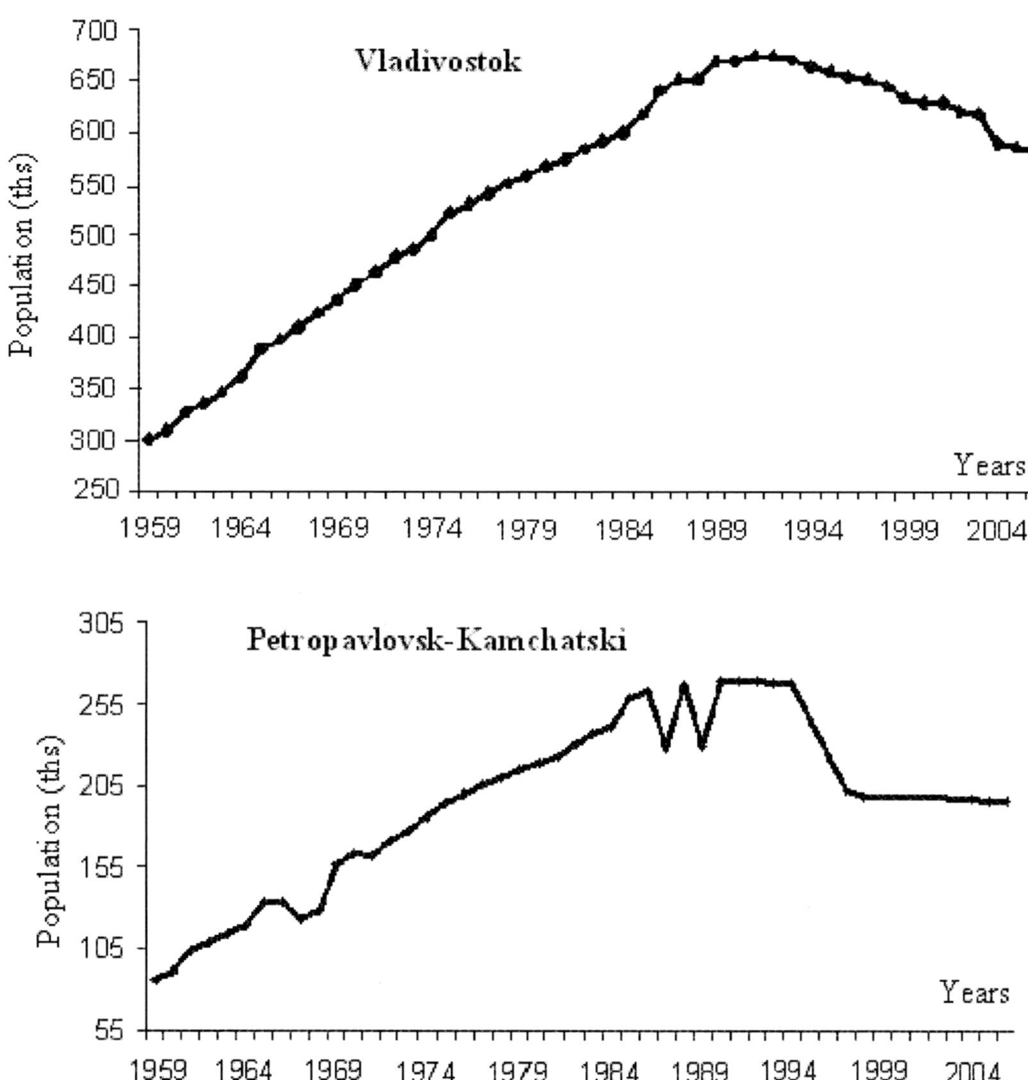

Figure 1. Population dynamics in Vladivostok and Petropavlovsk-Kamchatski.

The cutting down or completely stopping of government order led to strengthening of social tension. Having a lack of sufficient internal potential, small towns are losing 'replenishment' from the rural population. For a period of neoliberal market reforms the population increase was observed in two small cities of the Far East only. A number of small cities with definite specialization remain in stable condition on account of strategic, politic and economic prerequisites. About ten towns by the moment can be considered as points of stability that, under certain conditions, are able rapidly to restore their former population and became starting-points of growth in new conditions of economic development.

The graphic analysis of small cities dynamics revealed that small towns, with population up to 40 thousand, were exposed to the most considerable changes [7]. Approximately one third of towns with population less than 30 thousand returned to the post-war state level, that can mean only one thing – it is the present degradation of the existing long time system of the population distribution. Such cities transform into the category of rural settlements.

Especially great changes in urban population distribution took place in Sakhalin that was distinguished by a large quantity of small towns and particularly those situated in seaside areas (Figure 3).

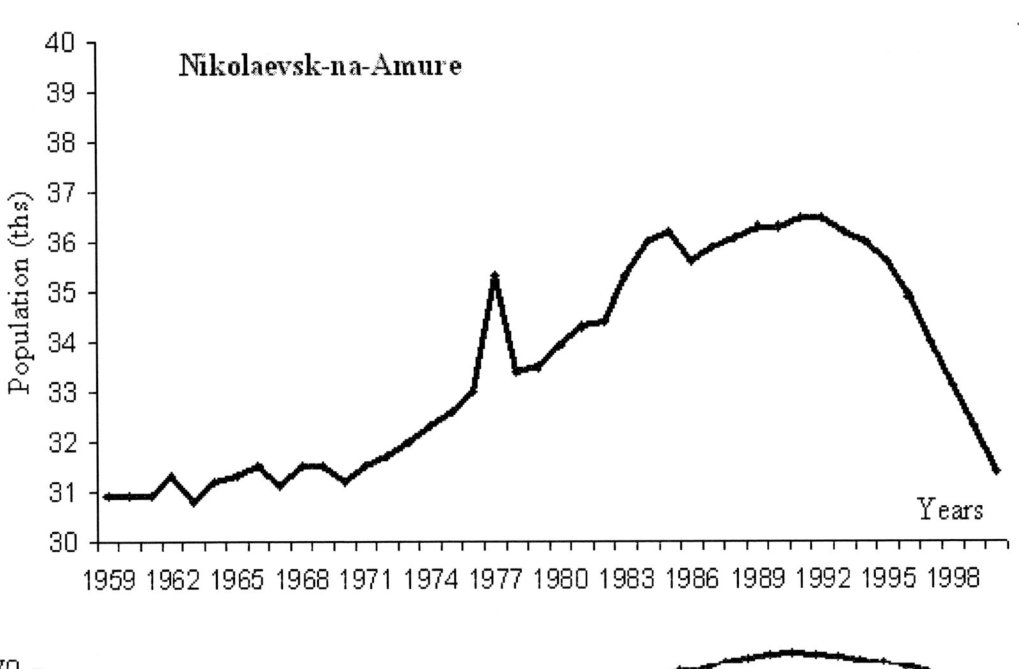

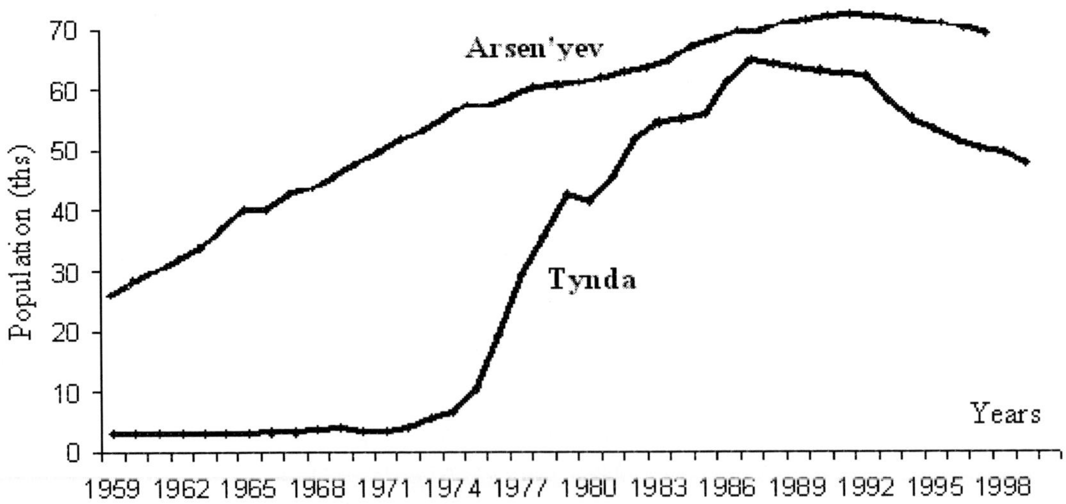

Figure 2. Population dynamics in Arsen'yev and Tynda.

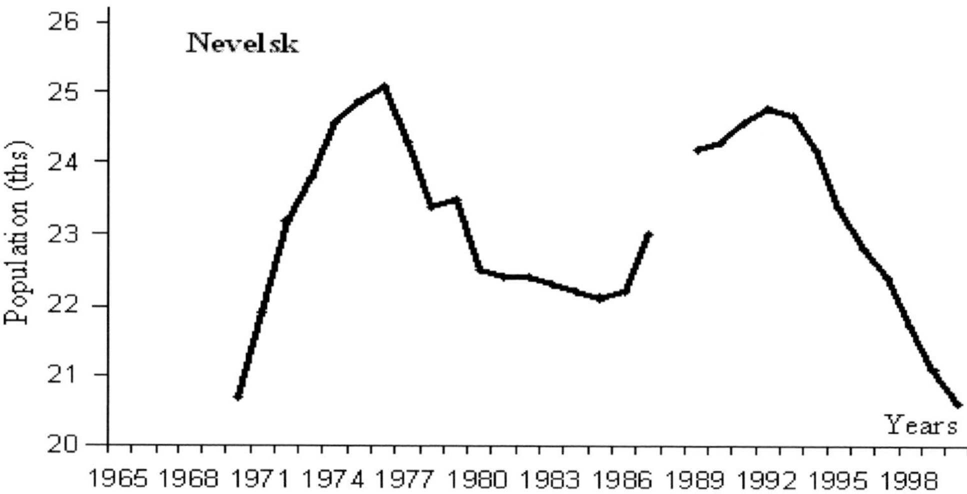

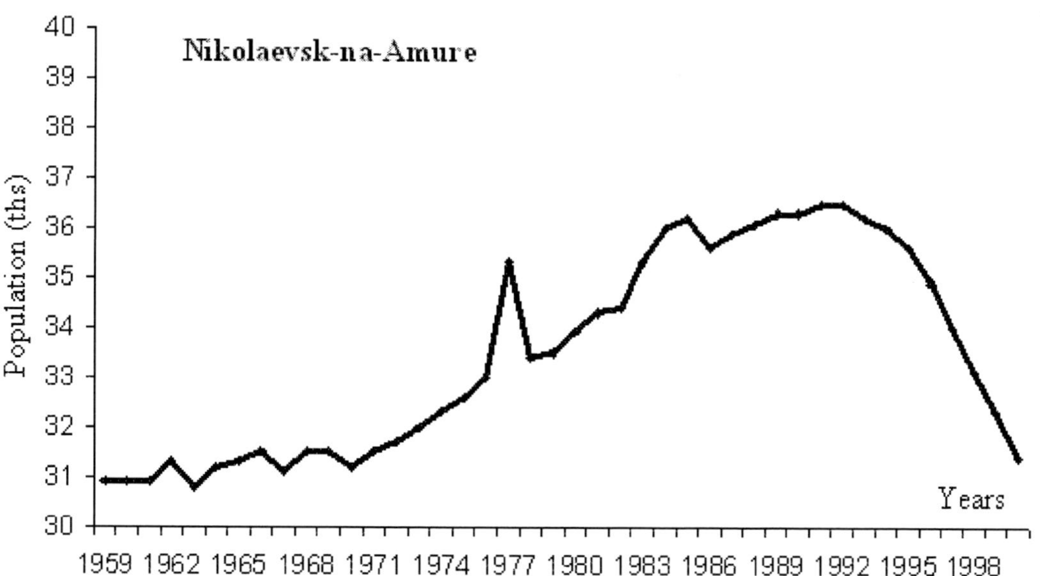

Figure 3. Population dynamics in Nevelsk and Nikolaevsk-na-Amure.

In the dynamics of small cities many things depend on economic and geographic location: nearness of traffic centers, frontier, latitudinal position inside the territory of federal district. For example, Anadyr city is the point of stability on the Chukotka, where population gradually is moving from other settlements of the Chukotski autonomous region.

The dynamics of urban-type community. The urban village, concerned with mining operations enterprises, became a more prevalent form of urban settling in the Far East. Economic reform resulted in reduction of some works, a shutdown of unprofitable enterprises. Found rather stable were the system of urban villages in Primorskiy Kray and Evreyskiy autonomous region, located mainly along the railway. Quite peculiar population dynamics are found in settlements situated close to the frontier with China (customs points). In other subjects of the federation, the population declined 20-40%.

In Primorskiy Kray the urban villages belonging to a fuel and energy complex (power station, coal production) have conditionally stable population dynamics as much as functioning of coal production enterprises permit. Part of the urban villages fall into a critical situation that resulted from a shutdown of coal mines they supply with labour. The settlements belonging to the mineral resource industry failed to keep population as well (Figure 4).

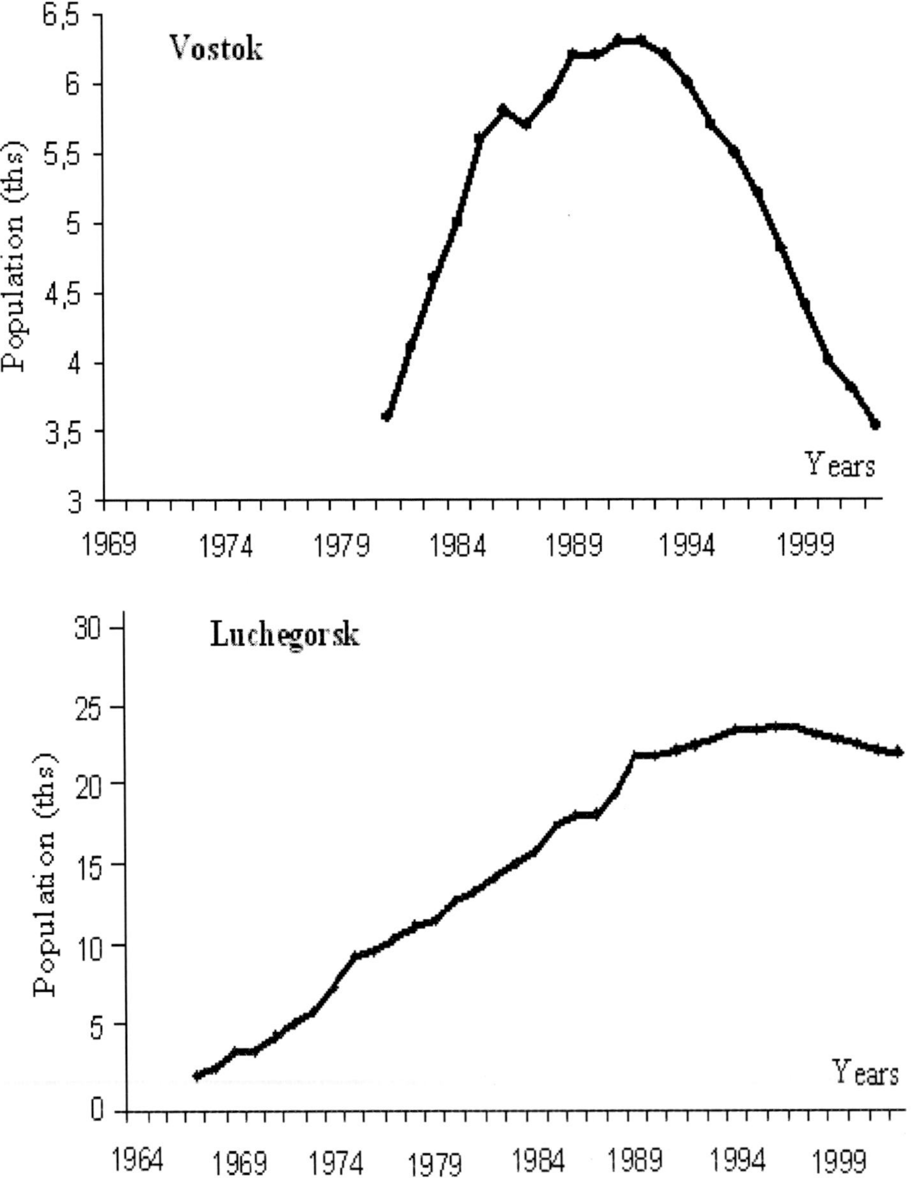

Figure 4. Population dynamics in Vostok and Luchegorsk.

The coastal settlements make up a separate group. Near-shore zone of Pacific littoral of Far East region was remarkable, enviable for dynamism of population upsurge. At the present time it is noted for two types of the population dynamics: stable growth and stable decline of population. Intensive growth is more visible in settlements with the supposed industrial growth. In analysis of population distribution it was considered, as a rule, that differences in socio-demographic processes (especially in the form where these differences are reflected by standard statistical indexes) depend on the population of a city.

Analysis of the same statistical data that was conducted by our method has shown that the conditionality of forming urban villages by territorial organization of national enterprise is losing its importance.

It turned out, that population dynamics in large and small cities have a number of similar features. The differences are taking on universal nature falling into a simple pattern – positives and negatives of living conditions in comparable cities. There are two main priorities, that effect for choice of habitation territory: job availability and transportation network in the city and its suburb.

Social polarization, defined by potential of survival and development is taking place today. There are two sharply defined groups of cities - the first one is a relatively viable part that is able to reorient its economic activity (in the first turn it is frontier cities) and the second group is degenerative cities with poor territorial population mobility.

The demographic situation is stable in those urban villages that have concentrated in themselves key industries, oriented either on federal level or originality resource. In support of dependence of the Far East from government decree we can say that the most important for region investment projects such as the railway bridge over Amur river at Khabarovsk, gas pipeline Komsomolsk-na-Amure – Amursk – Solnechny, new facility for extraction of lead-ore in Primorskiy Kray, new terminal in Vladivostok seaport, oil and gas wells on Sakhalin and so on were realized with the help of government investment (Figure 4).

Thus, using computer-based technologies, with the simple method of the visualization of population dynamics in cities, the authors could receive results that allowed in the shortest possible time to show the change rate in population dynamics. In keeping the positive trend in development of far-eastern cities, two factors became more important: nearness of frontier or military-industrial specialization, extremely necessary for the nation.

In the condition of the Far East integration into APR (Asia Pacific Region), the population distribution in the region would be considered by this time in context of spatial development of Far-East Asia as a whole. It is raising the role of global (macro) and local (micro) factors that define the living conditions of population. The contrasts between large cities and peripherals become stronger. The conception of "city- businessman", who earns one's living developing the most profitable projects, is getting more popular. Closely situated cities have indisputable advantages – they can supplement each other. Good examples of such associated cities are Vladivostok and Artem, having large agglomerate potential for setting up large-scale megapolis on south of Primorskiy Kray. The cities like Khabarovsk, Nakhodka, Komsomolsk-na-Amure are also promising on this point.

Stable development of most cities will depend on development of production, oriented to local and regional markets, works of enterprises of small business on processing of local farming raw material, production of the manufactured goods, for internal consumption, being in demand for customers.

The market mechanisms are not always able to ensure effective development and high economic growth. In order to mount to high rate of economic growth it requires coordination of efforts of both government and private capital. The constructive approach to the issue lies in confessing the presence of failure in state and market methods and in the search of conditions when cooperation of government and private business on various levels will give the best result in improving the demographic situation in the region.

REFERENCES

[1]   Sidorkina Z.I., Tsitsiashvili G.Sh. Using of System analysis in demographic researches. *Vestnik DVO RAN*, 2006. № 6, pp. 129-132. (In Russian).

[2]   Belskaya E.E. Small cities; Socio-economic and demographic problems and development perspectives. Abstracts of Ph.D. thesis Irkutsk, 2005. 25 p. (In Russian).

[3]   Tsitsiashvili G.Sh., Sidorkina Z.I. Small cities in the system of population distribution on the Far East:stability problems/ Perspectives of business development in small cities of the Far East: article collection/ edited by professor Osipov V.A. Vladivostok. Publisher DVGTU, 2007, pp. 24-31. (In Russian).

[4]   Baklanov P.Y. *The Far East Region of Russia: problems and premises of stable development.* Vladivostok: Dalnauka, 2001. 29 p. (In Russian).

Translated by Pavel Sidorkin (RoKcs/Vladivostok).

Chapter 7

EXPLORATION OF VARIABILITY IN THE ABOVE-EARTH AIR TEMPERATURE OVER THE FAR EAST REGIONS BY THE METHOD OF RESIDUAL VARIABILITY OF TEMPORAL ROW

T.A. Shatilina[1],, G.Sh.Tsitsiashvili[2],** and T.V. Radchenkova***

* Pacific Research Fisheries Center, Vladivostok, Russia
** Institute of Applied Mathematics, FEB RAS, Vladivostok, Russia

ABSTRACT

Regional climatic alterations in the Far East temperature of air were researched in the previous works [1, 3]. These authors estimated some tendency in the air temperature run over a century in the Asia-Pacific region. Intra-annual features in the trend changes and comparative analysis along separated stations, including shorter periods, were scarcely investigated. Nonetheless, under analysis of the century run of temperature the different periods are evident being distinctive by their tendency in alteration, and the significant trend attracts attention arising over certain seasons.

To analyze peculiarities in the alteration of above-earth air temperature above the Far East regions there was chosen the period from 1976 to 2005 years. This period is characterized by the most intense warming [2, 4].

The main purpose of the work is to find seasonal peculiarities in the tendency of air temperature alteration over various climatic zones of the Far East during the period of the most intensive warming by the method taking into account a residual changeability of temporal row.

---

[1] shatilina@tinro.ru.
[2] guram@iam.dvo.ru.

1. DATA AND METHOD OF EXPLORATION

Data on the temperature of air during 1948-2005 were taken from archive site www.giss.nasa.gov/data/update/gistemp/station_data/ and FERHRI for 17 stations of different climatic zones located at the Far East (Figure 1).

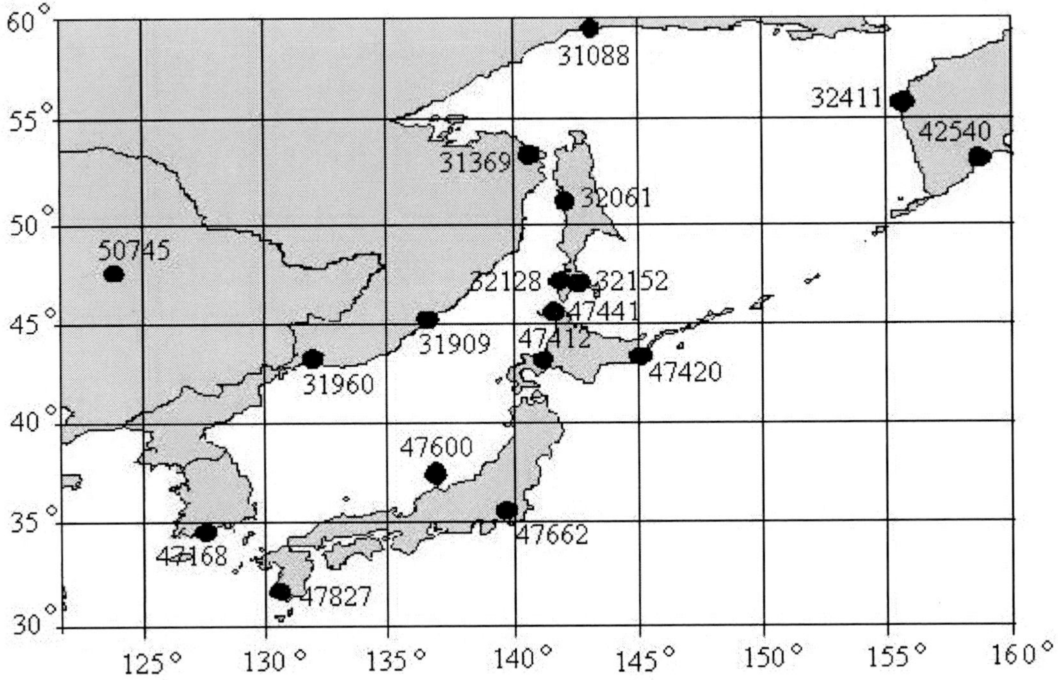

Figure 1. Location of hydrometeorological stations: 31088- Okhotsk, 32411 – Icha, 32540- Petropavlovsk-Kamchatsky, 31369 – Nikolaevsk-on-Amur, 32061- Aleksandrovsk Sakhalinsky, 50745 – Tsitsihar, 31909 – Terney, 31128- Kholmsk, 32152 – South Sakhalinsk, 47441- Wakkanai, 31960 – Vladivostok, 47412 – Sapporo, 47420- Nemuro, 47168 –Yeosu, 47600 – Wadsima, 47662 –Tokyo, 47827 – Kagoshima.

In the present work according to current temperature survey   using method of the best quadric, there is built the estimation of a, b and c, d coefficients in the equation $T = ct + d$ linear trend, that is minimizing function

$$s(c,d) = \sqrt{\frac{1}{n} \sum_{i=1}^{n} \left(T(t) - (ct + d)\right)^2}.$$

Here a characterizes the temperature alteration during time unit (1 year). Then, the quadratic mean deflection of current temperature from linear trend of equation $T = at + b$ (residual variability of temporal row) will be determined by $s = s(a, b)$ which was calculated under a, b optimal magnitudes. In our case $T = \dfrac{temper.}{1°C}$, $t = \dfrac{time}{1 year}$, we have temperature and time without size. So, a, b - are magnitudes that have no dimension. Then n – is also the magnitude without size, that features the number of observation years. Alongside with determination of sizeless a, s there is calculated their ratio s/a. Magnitude s/a (or reciprocal a/s) features statistic amount of linear trend coefficient (particularly, confidence interval of this coefficient). For some stations the magnitude of a/s is so great, that it means statistic value of trend is small. It shows that residual variability prevails over the trend alteration of temperature. Therefore the study of residual variability for some stations is interesting for comparison. If to compare there should be underlined, that statistic volume of the trend calculated on large number of stations is significant, so that is why residual variability is moved to the secondary meaning. Computation was accomplished in Excel program.

2. PRINCIPAL RESULTS OF RESEARCH

Figure 2 exhibits magnitudes of the linear trend coefficient a for 1976-2005 years in comparison to 1948-1975 years.

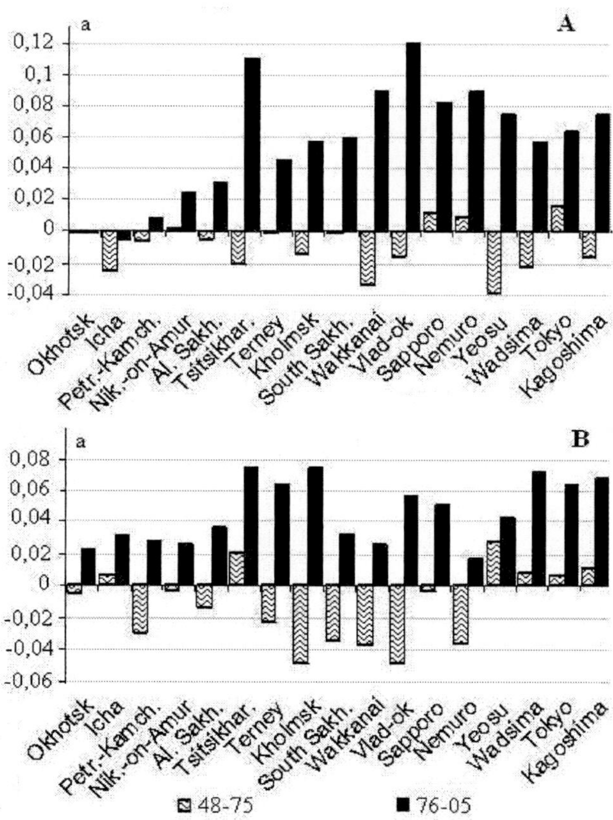

Figure 2. Magnitudes of the linear trend coefficient a for the station to row in latitude from North to South: A – February, B– September for: 1948-1975 and 1976-2005 periods.

In January and February 1976-2005 all the Far East stations had positive coefficients of linear trend. At the station of Tsitsikhar and Vladivostok the magnitudes of a exceeded 0.07. During the 1948-1975 period these coefficients of linear trend were almost 0 often with negative sign. The most intensive growth of above-earth air temperature was surveyed at the stations in the temperate and south zones.

In May and June the tendency to growth of the linear trend coefficient during 1976-2005 is preserved in comparison with the 1948-1975 period. Maximum increase of temperature was watched at the station on Tsitsikhar (0.08), Icha (0.07), South-Sakhalinsk (0.07).

In September and October, the Far East stations1 had such coefficients of linear trend on the temporal sections of 948-1975 and 1976-2005 periods, which were essentially different (Fig. 2 B). During 1976-2005 there was detected a group of stations, where the coefficients of linear trend were high, namely: Tsitsikhar, Terney, Kholmsk (stations of temperate latitudes) and Southern stations: Wadsima, Tokyo, Kagoshima, where the magnitude a was equal 0.07-0.08.

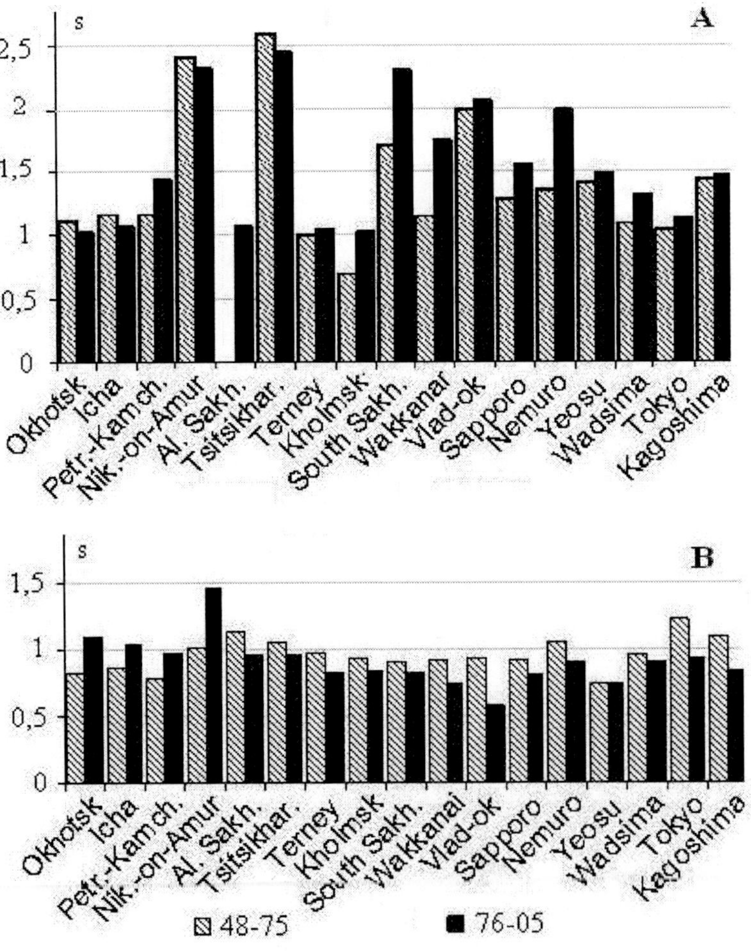

Figure 3. Magnitudes of residual variability in temporal row S at the Far Eastern stations in February (A), September (B).

In January and in February the increase of residual variability was observed at the majority of stations during the 1976-2005 period compared to 1948-1975. Decrease s was noticed only in Nothern stations, namely, Okhotsk, Icha, Petropavlovsk-Kamchatsky. The biggest residual variability was discovered in the stations of Kholmsk, South-Sakhslinsk, Nemuro, Wadsima.

During June from 1976-2005 on the majority of stations there was underlined the tendency to a rise of residual variability of air temperature being compared with an earlier period.

On the Far East Northern stations the changeability of temperature in September during the late year's period was more, then during the earlier period (Figure 3 B). This ratio was reciprocal on the rest of the stations.

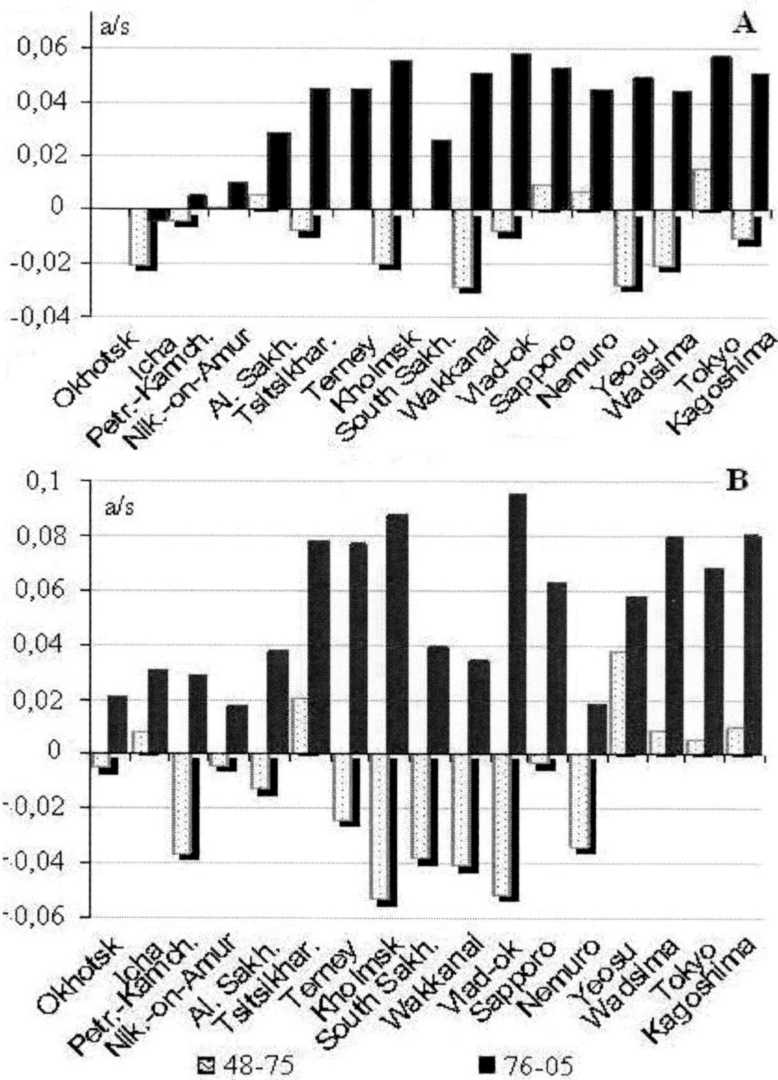

Figure 4. Magnitudes of statistic significance of air temperature trend a/s at the Far East stations in February (A) and in September (B) for the 1976-2005 period.

The trend alteration in air temperature is intensified at the majority of stations in September 1976-2005 being distinctive from the winter period. It is an essential difference from the winter period.

In order to confirm the congruence of statistic calculations with natural data, Figure 5 (A-Г) displayed the anomaly run in the air temperature of some Far Eastern stations, where the most essential features were disclosed in changes of the above-earth air temperature during winter and autumn period of 1976-2005 years.

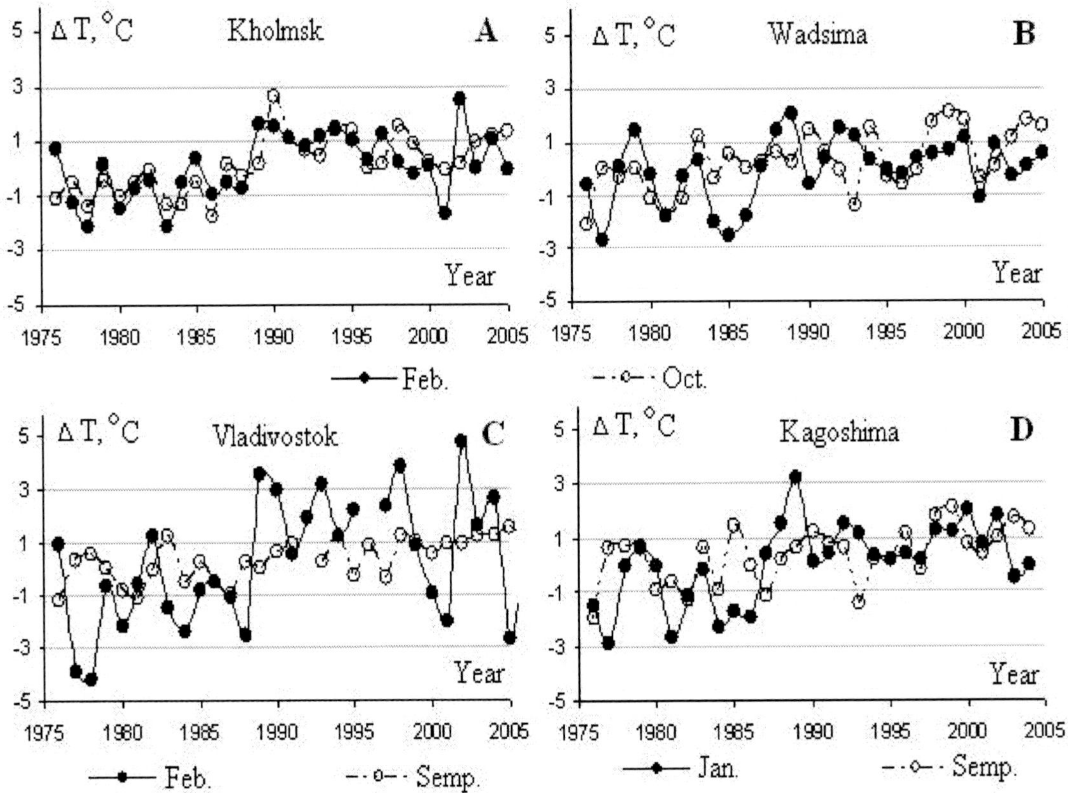

Figure 5. Anomalies of air temperature at the Far East stations (deviation of the average temperature for basis period 1971-2000 years) during 1976-2005 years, namely: A –Kholmsk, B – Wadsima, C – Vladivostok, D – Kagoshima.

It is evident, that the rise of air temperature was equal to 1.49°C near the coast of Western Sakhalin (Kholmsk) in February from the period of 1976-1988 to the span of 1989-2005, and in October it was equal to – 1.62°C. At this station in February the magnitude of trend was equal to 0.06, but in autumn it increased to 0.09. At the South of Primorye (Vladivostok) in February the air temperature from the period to period rose by 2.9° C, and in September only by 0.95°C. For Vladivostok the statistic magnitude of the trend in February reduced till 0.06 due to increase of residual variability, but in September it became higher to 0.1. At the station Wadsima the air temperature fluctuations are comparable in winter and in autumn, temperature rose by 1° C in both seasons, but magnitude of trend was higher in September (from 0.04 in February till 0.08 in September). At the South of Japan (Kagoshima) in January the air temperature rose by 1.7° C, and in September by 1°C. Magnitude of the trend was increased in September at this station too if to compare with the winter month (from 0.05 to 0.08).

CONCLUSION

Thus, surveys show the growth of above-earth temperature of air in 1976-2005 compared to the prior period of 1948-1975 years at the most number of stations situated in diverse climatal zones of the Far East. Maximum increase of air temperature during winter was watched at the North-East of China and above Primorye. During autumn the maximum increase of air temperature was marred along the coast of Primorye, Sakhalin Island and Japan.

Residual variability augment of the air temperature during the winter period and decrease in the autumn period were surveyed during the 1976-years span.

Statistical value of the temperature trend that was identified applying the original a/s, coefficient and is being magnified in the autumn span and is being lowered in the winter span.

The statistical analysis results of air temperature temporal row tally with the natural data and supplement our knowledge of regional alterations of climatal parameters.

REFERENCES

[1]   Gayko L.A. Tendency of water and air temperature fluctuations in the coastal zone of the north-western Japan Sea. *Far Eastern seas of Russia*. Book 1: Oceanologic research. Moscow: Nauka, 2007. Pp. 307-332. (In Russian).

[2]   Gruza G.V., Ran'kova Eh. Ya. Finding of climatic change: condition, change, extremely. *Meteorology and Hydrology*, 2004. № 4. Pp. 50-67. (In Russian) .

[3]   Ponomarev V.I., Kaplunenko D.D., Dmitrieva E. V., Novorotsky P.V. Climate variations in the northern part of Asia-Pacific region. *Far Eastern seas of Russia*. Book 1: Oceanologic research. Moscow: Nauka, 2007. Pp. 17-48. (In Russian).

[4]   Folland C. K., Rayner N. A., Brown S. J., et al. Global temperature change and its uncertainties since 1861. *Geophys. Res. Lett.*, 2001. Vol. 28, № 13. Pp. 2621- 2624.

Translated by Vera Kochetova (Far Eastern State Technical University).

In: Efficient Algorithms of Time Series Processing...   ISBN 978-1-60692-062-6
Editor: G. Sh. Tsitsiashvili   © 2009 Nova Science Publishers, Inc.

Chapter 8

ESTIMATES OF VARIANCES IN TIME SERIES STATISTICS

G.Sh. Tsitsiashvili [1]

Institute of Applied Mathematics, FEB RAS, Vladivostok, Russia

A problem of an estimation of a random deviation between observations and a regression function is considered. This problem originated from an analysis of time series of over ground air dynamics. These observations attract large interest in connection with the global climate warming phenomenon. So a problem of an estimation of these fluctuations variances is actual now.

In classical statistics a problem of variance estimation is solved by an empirical variance. But even for this widely used statistic, it is complicated to calculate its own variance. To calculate a variance of an estimate for a variance of a deviation from a polynomial regression function is much more difficult. Nevertheless it is possible to solve this problem if observations are made in integer-valued points.

This problem is solved in two steps. At the first step an analog of empirical variance is constructed so that it is possible to calculate its own variance. At the second step for a regression function, represented as a polynomial of integer-valued argument, a special recurrent procedure is constructed. This procedure transforms random observations of the polynomial into a sequence of independent and identically distributed random variables. Then it is possible to use results of the first step. This algorithm is realized without an estimation of the regression function coefficients. All these constructions may be spread to a multidimensional case easily. In this case an estimation of a variance is to be replaced by an estimation of a covariance's matrix with an estimation of its elements variances. This generalization is initiated by applied problems of mathematical geodesy.

---

[1] guram@iam.dvo.ru.

1. MODIFIED EMPIRICAL VARIANCE

Suppose that $x_1, x_2, \ldots$ is the sequence of independent and identically distributed random variables (i.i.d.r.v`s) with the common distribution function (d.f.) $F(t)$. Denote

$$\mathbf{E} x_1 = \int_{-\infty}^{\infty} t \, dF(t) = a, \quad \mathbf{Var}\, x_1 = \mathbf{E}(x_1 - a)^2 = b < \infty.$$

Usual estimates of a, b are the empirical expectation and the empirical variance [1]:

$$\hat{a}_n = \frac{1}{n} \sum_{i=1}^{n} x_i, \quad \hat{b}_n = \frac{1}{n-1} \sum_{i=1}^{n} (x_i - \hat{a}_n)^2$$

which are unbiased. The empirical expectation $\hat{a}_n$ has the variance $\mathbf{Var}\, \hat{a}_n = b/n$. But a calculation of $\hat{b}_n$ variance is a sufficiently complicated procedure. So in this section we consider the following unbiased estimate of b for which a calculation of its variance is apparent.

Introduce i.i.d.r. v`s $z_1 = x_1 - x_2$, $z_2 = x_4 - x_3, \ldots$ satisfying the equalities

$$\mathbf{E}\, z_1 = 0, \quad \mathbf{Var}\, z_1 = 2b. \tag{1.1}$$

Using r.v`s $x_1, \ldots, x_{2n}$ define the estimate b_n' of b by the formulas:

$$b_n' = \frac{1}{2n} \sum_{i=1}^{n} z_i^2.$$

From the equalities (1.1) obtain: $\mathbf{E}\, b_n' = b$,

$$\mathbf{Var}\, b_n' = \frac{1}{n} \mathbf{E}\left[\frac{z_1^2 - 2b}{2} \right]^2 = \frac{1}{4n} \mathbf{E}\left[z_1^2 - 2b \right]^2.$$

For r.v. u define $\bar{u} = u - \mathbf{E}\, u$ then $b = \mathbf{E}\, \bar{x}_1^2$. Denote $d = \mathbf{E}\, \bar{x}_1^{2^2}$ and obtain the equalities

$$\mathbf{Var}\, b_n' = \frac{1}{4n} \mathbf{E}\left[(\bar{x}_2 - \bar{x}_1)^2 - 2b \right]^2 = \frac{1}{4n} \mathbf{E}\left(\bar{x}_1^{2^2} + \bar{x}_2^{2^2} - 2\bar{x}_1 \bar{x}_2 \right)^2 =$$

$$= \frac{1}{4n} \mathbf{E} \left[\overline{x_1^2}^2 + \overline{x_2^2}^2 + 4\overline{x_1^2}\,\overline{x_2^2} \right] = \frac{1}{4n} \left[2\mathbf{E}\overline{x_1^2}^2 + 4\left(\mathbf{E}\overline{x_1^2}\right)^2 \right] = \frac{d+2b^2}{2n}.$$

So

$$\mathbf{E}\,b_n' = b, \quad \mathbf{Var}\,b_n' = \frac{d+2b^2}{N}, \quad N = 2n. \tag{1.2}$$

Assume now that i.i.d.r.v`s $x_1, x_2, \ldots$ have normal distribution then b_n' is the likelihood estimate for the sequence $z_1, \ldots, z_n$. In an accordance with well known formulas [2] for moments of normal distribution

$$d = 2b^2 \tag{1.3}$$

and consequently from (1.2) we obtain

$$\mathbf{Var}\,b_n' = \frac{4b^2}{N}. \tag{1.4}$$

2. VARIATION OF DEVIATION FROM POLYNOMIAL REGRESSION FUNCTION

Suppose that the random sequence $y_{1,0}, y_{2,0}, \ldots$ satisfies the equalities $y_{i,0} = P_m(i) + \varepsilon_{i,0}$ where $P_m(i) = \sum_{j=0}^{m} i^j p_{m,j}$ is the polynom of the degree m, $\varepsilon_{1,0}, \varepsilon_{2,0}, \ldots$ are i.i.d.r.v`s satisfying the equalities

$$\mathbf{E}\varepsilon_{1,0} = 0, \quad \mathbf{Var}\,\varepsilon_{1,0} = \mathbf{E}\,\varepsilon_{1,0}^2 = \beta_0, \quad \mathbf{E}\,\overline{\varepsilon_{1,0}^2}^2 = f_0.$$

Define recurrently the random sequences

$$\left\{ y_{1,1}, y_{2,1}, \ldots \right\}, \; \left\{ y_{1,2}, y_{2,2}, \ldots \right\}, \ldots, \left\{ y_{1,m}, y_{2,m}, \ldots \right\}$$

as follows

$$y_{i,k+1} = y_{2i,k} - y_{2i-1,k}, \quad 1 \le i, \; k = 0, \ldots, m-1.$$

By the definition

$$y_{i,k+1} = P_{m-k-1}(i) + \varepsilon_{i,k+1}, \ \varepsilon_{i,k+1} = \varepsilon_{2i,k} - \varepsilon_{2i-1,k},$$

$$P_{m-k-1}(i) = P_{m-k}(2i) - P_{m-k}(2i-1), \ i \geq 1, \ k = 0,\ldots,m-1.$$

Using the equality $a^r - b^r = (a-b)\left(a^{r-1} + a^{r-2}b + \ldots + b^{r-1}\right)$ which is true for arbitrary natural number r obtain

$$P_{m-k-1}(i) = P_{m-k}(2i) - P_{m-k}(2i-1) = \sum_{j=0}^{m-k}(2i)^j p_{m-k,j} - \sum_{j=0}^{m-k}(2i-1)^j p_{m-k,j} =$$

$$= \sum_{j=1}^{m-k} p_{m-k,j} \sum_{t=0}^{j-1}(2i)^t (2i-1)^{j-1-t}.$$

Then from the binomial theorem it is easy to obtain that the function $P_{m-k-1}(i)$ may be represented as the polynomial of the degree $m-k-1$ of integer argument i:

$$P_{m-k-1}(i) = \sum_{j=0}^{m-k-1} p_{m-k-1,j}\ i^j.$$

So $\mathbf{E}\,\varepsilon_{i,k+1} = 0$, $\mathbf{Var}\,\varepsilon_{1,k+1} = \beta_{k+1} = 2\beta_k$ then

$$\mathbf{E}\,\overline{\varepsilon_{1,k+1}^2}^2 = f_{k+1} = 2f_k + 4\beta_k^2 = 2^{k+1}f_0 + 4\sum_{i=0}^{k}2^{k-i}\beta_i^2 = 2^{k+1}f_0 + 4\beta_0^2\sum_{i=0}^{k}2^{k+i} =$$

$$= 2^{k+1}f_0 + 4\beta_0^2 2^k\left(2^{k+1}-1\right) = 2^{k+1}\left(f_0 + 2\beta_0^2\left(2^{k+1}-1\right)\right)$$

and consequently

$$\beta_m = 2^m\beta_0, \ f_m = 2^m\left(f_0 + 2\beta_0^2\left(2^m-1\right)\right), \ m \geq 1.$$

Define now $x_i = y_{m,i}$, $i \geq 1$, $b = \beta_m$, $d = f_m$ and construct by the sample $x_1,\ldots,x_{2n}$ the estimate $b'_{n,m} = b'_n/2^m$ of β_0: $\mathbf{E}\,b'_{n,m} = \beta_0$,

$$\mathbf{Var}\,b'_{n,m} = \frac{d+2b^2}{2^{2m+1}n} = \frac{f_m + 2\beta_m^2}{2^{2m+1}n} = \frac{2^m\left(f_0 + 2\beta_0^2\left(2^m-1\right)\right) + 2^{2m+1}\beta_0^2}{2^{2m+1}n} = \frac{f_0 + \beta_0^2\left(2^{m+2}-2\right)}{2^{2m+1}n}$$

.

If i.i.d.r.v`s $\varepsilon_{i,0}$, $i \geq 1$, have normal d.f., then $f_0 = 2\beta_0^2$ and so $\mathbf{Var}\, b'_{n,m} = \dfrac{2\beta_0^2}{n}$. To construct the estimate $b'_{n,m}$ it is necessary to have the sample $y_{1,0},..,y_{N,0}$ consisting of $N = 2^{m+1}n$ members.

3. MODIFIED EMPIRICAL COVARIANCE

Consider i.i.d.r. vectors

$$Z = \left(z_1,...,z_m\right),\ Z_1 = \left(z_{11},...,z_{1m}\right),\ Z_2 = \left(z_{21},...,z_{2m}\right),...$$

Suppose that $\mathbf{E}\, z_1 = ... = \mathbf{E}\, z_m = 0$. Random vectors $Z, Z_1, Z_2,...$ have common multidimensional distribution defined by i.i.d.r. vectors $\left(e_1,...,e_m\right)$, $\left(e_{11},...,e_{1m}\right)$, $\left(e_{21},...,e_{2m}\right),...$ as follows

$$z_j = \sum_{t=1}^m a_{jt}\, e_t\,,\ z_{ij} = \sum_{t=1}^m a_{jt}\, e_{it}\,,\ i \geq 1\,,\ 1 \leq j \leq m\,.$$

The random vector $\left(e_1,...,e_m\right)$ consists of i.i.d.r. components with zero mathematical expectation, single variance and finite fourth moment μ, the matrix $\left\|a_{jt}\right\|_{j,\,t=1}^m$ consists of real numbers.

The covariance $\mathbf{cov}\left(z_t, z_s\right) = c\left(t,s\right)$ satisfies the equality

$$c\left(t,s\right) = \sum_{k=1}^m a_{tk}\, a_{sk}\,.$$

An unbiased estimate of the covariance $c\left(t,s\right)$ is

$$\hat{c}_n\left(t,s\right) = \frac{1}{n}\sum_{i=1}^n z_{it}\, z_{is}\,,\ \mathbf{Var}\, \hat{c}_n\left(t,s\right) = \frac{1}{n}\mathbf{Var}\, z_t z_s$$

where

$$\mathbf{Var}\, z_t z_s = \mathbf{Var} \sum_{j,k=1}^m a_{tj}\, a_{sk}\, e_j\, e_k = b\left(t,s\right) - c^2\left(t,s\right),$$

$$b(t,s) = \mathbf{E}\left[\sum_{j=1}^{m}\sum_{k=1}^{m} a_{tj}\, a_{sk}\, e_j\, e_k\right]^2 = \sum_{j=1}^{m}\sum_{k=1}^{m}\sum_{r=1}^{m}\sum_{l=1}^{m} a_{tj}\, a_{tr}\, a_{sk}\, a_{sl}\, \mathbf{E}\, e_j\, e_r\, e_k\, e_l \,.$$

From the random vector $(e_1,...,e_m)$ definition obtain

$$b(t,s) = \mu \sum_{j=1}^{m} a_{tj}^2\, a_{sj}^2 + \sum_{1\le j,\, r\le m,\, j\ne r} a_{tj}\, a_{tr}\, a_{sj}\, a_{sr} + \sum_{1\le j,\, k\le m,\, j\ne k} a_{tj}^2\, a_{sk}^2 + \sum_{1\le j,\, k\le m,\, j\ne k} a_{tj}\, a_{tk}\, a_{sk}\, a_{sj}$$

.

Then

$$\mathbf{Var}\,\hat{c}_n(t,s) = \frac{(\mu-3)d(t,s) + c^2(t,s) + c(t,t)c(s,s)}{n},$$

$$d(t,s) = \sum_{j=1}^{m} a_{tj}^2\, a_{sj}^2 \le c(t,t)c(s,s).$$

If the components of the random vector $(e_1,...,e_m)$ have normal distribution then from (1.3) $\mu = 3$ and so

$$\mathbf{Var}\,\hat{c}_n(t,s) = \frac{c^2(t,s) + c(t,t)c(s,s)}{n}.$$

Remark 1. A suggested technique of a covariance estimate and a calculation of its variance by means of previous section formulas may be generalized to deviations of a multidimensional regression function which is a polynomial function of integer multidimensional argument.

REFERENCES

[1]　　Rozanov Yu.A. *Probability theory, random processes and mathematical statistics.* M.: Science, 1971. 318 p. (In Russian).

[2]　　Orlov A.I. *Mathematics of chance: Probability and statistics - main facts*: Tutorial. M.: MZ-Press, 2004. 110 p. (In Russian).

Translated by Guram Tsitsiashvili (Institute of Applied Mathematics, FEB RAS).

INDEX